The Kaizen Revolution

The Kaizen Revolution

How To Use Kaizen Events
To Double Your Profits

Michael D. Regan
with
Mark Slattery

Holden Press

The Kaizen Revolution

How To Use Kaizen Events
To Double Your Profits

By Michael D. Regan
with
Mark Slattery

Published by:
Holden Press
Raleigh, North Carolina
(888)910-8326

ISBN
0-9663549-7-4

Acknowledgements

First I acknowledge Almighty God, without whom I could not exist, let alone write a book.

Thank you, Stacey. You are a wonderful wife and mother, and an excellent sounding board. Thanks also to the rest of my family for their support and wisdom, and to my clients and colleagues for their thoughtful input.

Special thanks to George Foster for his unparalleled creativity and efficiency in designing my book covers, and to Mandi Stanley for her professional proofreading.

I also thank the following people for reviewing the manuscript and giving me thoughtful feedback: Mike Seleway, Tim Weilbaker, Scott Spears, Eric Floodeen, Fred Hardee, Peggy Saylor, Jody Barish, George Wells, Olivier Larue, Alan Fox, Ben Levitan, David Burke, Chuck Millard, Ana Tampanna.

About the Authors

Michael D. Regan, chairman of Everest Consulting Group, Inc., is a professional speaker and consultant. He has worked with a wide variety of companies and industries in 31 states and internationally, including DuPont, GlaxoWellcome, Alcatel, Nortel, Borg-Warner Automotive, Sumitomo Electric, BB&T Bank, and Carolina Medical Review. He holds an MBA from the University of Rochester. He is also the author of *The Journey To Teams: The New Approach to Breakthrough Business Performance*.

Mark Slattery, CEO of Everest Consulting Group, has spent 22 years helping organizations build teams and improve productivity. His clients have ranged from the Fortune 500 to small privately held companies, and have included manufacturers, banks, government agencies, health care organizations, utilities, and telecommunications firms. Mark was a contributor to the book *The Journey to Teams*, is a graduate of Merrimack College, and plays a mean guitar.

TABLE OF CONTENTS

Pronunciation:

The word "kaizen" is pronounced "kye-zen."
It rhymes with "pie-pen."

Note From the Author

Bite off more than you can chew, then chew it. Plan more than you can do, then do it.

— Anonymous

What are kaizen events, and can you really use them to double your profits? Kaizen is a Japanese word which means improvement. A kaizen event is a focused week-long improvement effort. A well-planned and expertly led event can produce one or more of the following results: lead time and work-in-process reduced by 80 percent, scrap reduced by 50 percent, and productivity increased by 30 percent. These improvements will help you simultaneously increase your market share and reduce your costs, resulting in dramatically better profits.

The ideas in this book are presented in the form of a story for ease of reading and maximum learning. Each chapter ends with a summary of key points for efficient review.

Dive into this book, dream about what you can accomplish, then go out and do it.

Michael D. Regan
Raleigh, NC
February 2000

Bad News

Thursday, April 8   9:44 a.m.

Cecil Langdon III, owner and CEO of Accurate Clock Company, crossed the street to West-Mart corporate headquarters in downtown Indianapolis. He felt sure West-Mart would ask for another five percent price reduction this year, as they had each of the previous two years.

Accurate Clock Company was the sole supplier of wall clocks to West-Mart, an account which represented nearly half of his company's sales. Recently, however, their demands were driving him into bankruptcy. He lost money on every clock he sold, and now they were going to ask him to lose more.

He walked through the front door and signed in with the receptionist. He had five minutes before his appointment. Sitting on a couch in the lobby, he realized that the battle to save his company had exhausted him and ravaged his life. He was 54 years old, 60 pounds overweight, and had not slept more than five hours a night for the past two years. Eighteen months ago, his wife had divorced him because he spent too much time at work. Nonetheless, his grandfather had started the company, and he did not want to be the one to end it.

"Cecil, come on in," said a voice to his left. Cecil looked up to see Ted Wittenburg, West-Mart's V.P. of purchasing, leaning into the lobby through the open interior door.

"Good morning," Cecil said as he collected his briefcase and stood awkwardly. They shook hands and Cecil followed him into a nearby conference room.

"I don't have a whole lot of time for our meeting this morning, but we'll wrap this up quickly," said Ted as he motioned for Cecil to sit. "Everything good in Fort Wayne these days?"

"To be honest with you, Ted," said Cecil, "things are getting really tight. We've cut your cost by 10 percent over the past two years. We're losing money and we need this year to catch up. Any more pressure from

2

West-Mart and we may go under."

"There won't be any more pressure from us."

"That's great, Ted."

"You don't understand, Cecil. We're dropping you."

"What?" asked Cecil in disbelief. He felt his stomach turning over and the blood rushing to his face.

"We've already signed a contract with an overseas supplier who can meet the delivery and quality goals we are asking for."

"But," protested Cecil, "we've busted our tails for you. We lost money all last year because of you..."

"You said yourself that you need this year to catch up," interrupted Ted, "and that you would go under if we put any more pressure on you. I believe that and I admit you've met the cost targets. The problem is that your poor quality and erratic delivery performance have cost us more in paperwork, reputation, and lost sales than your cost reductions have saved us."

"But we...," Cecil tried to interject.

"Cecil," Ted interrupted again firmly, "we're convinced that your company cannot meet our standards. You know that yourself. We've chosen another vendor and that's that. In fact, we're paying them more per unit in order to get better quality and delivery. I'll walk you out now."

Cecil felt dizzy. He tried to stand, but fell sideways,

hit the chair on his right, and landed heavily on the floor.

"You okay, Cecil?" asked Ted as he reached out a hand to help him stand up. "I'm sorry this had to happen. We're running a business here just like you are, and this is a decision we had to make."

"Right, right," mumbled Cecil as he maneuvered himself around Ted and down the short hallway to the lobby. He stumbled through the front doors and out onto the sidewalk.

His vision was blurred now, and he was having trouble breathing. He had sacrificed his health and his marriage to keep the family business alive, and now it was certain he had failed. As he made his way toward the crosswalk, a paralyzing pain seized the entire right side of his body, causing him to lurch forward and to his right. His body relaxed as his head hit the pavement.

The Prodigal Son

Saturday, April 10   3:25 p.m.

C. Thomas Langdon IV stood motionless in the rain as his father's body was lowered into the ground. This was the latest twist in a recent succession of events that had rocked his life. Until recently he had been doing very well financially as a bond trader for Goldstone Sherman, the Wall Street investment bank. Three months ago he had bet almost $50 million dollars of his firm's portfolio, plus his entire savings and as much money as he could borrow, on the cheap bonds of a struggling company. He thought for sure the company would turn itself around and make him and his firm a quick profit. Two weeks later the company declared bankruptcy, and

5

Goldstone Sherman fired Tom. A month after that, his fiancée left him, last week he lost his apartment, and two days ago his father died. So here he was back where he started in Fort Wayne, Indiana. As his father's only child, he had inherited the Accurate Clock Company, founded by his great-grandfather and run by Langdon men for 81 years. All he could think about now was selling it as fast as possible and using the money to reclaim his vastly more stimulating life in New York City.

*   *   *

Monday, April 12

Tom drove his BMW through the already crowded parking lot and pulled into his father's reserved spot to the right of the front entrance. As he stepped out of the car he slid a dark blue suit coat over his starched white shirt, then checked the time on his Rolex watch: 8:55 a.m. He stepped confidently onto the concrete walkway and pushed open the door. To his left was a mirror. He stopped and looked at himself, smoothed back his black hair and smiled. With his financial connections, he would be out of town in a week with all the money he needed.

"Tom," said an abrupt voice from behind him.

He turned to see old Roy Esterhaus, dressed in tan

slacks and a golf shirt, looking at him with barely concealed contempt. This new casual dress code was fine for Roy and manufacturing types, but Tom was not one of them. He was a businessman.

"Roy," Tom said, staring back at him. He did not bother to offer his hand. Roy was in his late sixties and had been chief engineer at Accurate Clock for 36 years. The older man had always been offended by Tom's lifelong and obvious disinterest in the family business.

That was too bad, wasn't it, thought Tom. Well, he was interested now and would continue to be until he turned this company into cash.

"Who's the controller around here these days?" Tom asked.

"Oh, that'll be a good one," Roy said. "Last door on the left before your father's office." He pointed, then turned his back and walked down the hall toward the manufacturing floor.

"What do you mean?" demanded Tom. Roy shook his head and kept walking.

Tom walked down the hallway and noticed that there had not been a new coat of paint applied to the walls since his last visit as a high school senior nearly 15 years ago. He arrived at the correct office, started to knock on the open door, and froze. "Maria?" he asked, stunned.

The woman turned away from her conversation with

a man in a dark suit. "Sorry about your father, Tom," she said.

"Thanks," he stammered. He had dated Maria Vasquez from his first year of high school until the beginning of his second year at Harvard. She was at Indiana State at the time and the distance caused them to drift apart. He thought of her often, but they hadn't talked in years. She looked great. No ring on her finger.

"I saw you at the funeral," she said, "but I knew you were preoccupied and that I'd see you here today."

"So you work here now," he concluded.

"Not exactly," she said. "I'm a manager with the accounting firm Gatewood & Wooten. Based on my advice, your father fired the controller last week. I stepped in to act as temporary controller and to help him figure out the financial situation here."

"Then you're the one I need to talk to," he said, getting down to business. "I'm going to sell this company, and I need to know what it's worth."

She opened her mouth to speak, then frowned. "Tom, this is Mortimer Stern from Indiana First Trust Bank," she said, gesturing to the man beside her. "Mortimer, this is Tom Langdon, Cecil's son."

"This isn't your company anymore," Mortimer said to Tom. "Accurate Clock Company owes my bank $11 million, and your father missed last month's payment.

According to our contract, we now can sell this company to recover our money, and that's what I intend to do."

"Because of one missed payment?" Tom asked angrily. "Frank Fowler is the president of Indiana First Trust and my father's been doing business with him for years. Frank wouldn't do this to him."

"You're right," said Mortimer, "but now your father's gone and so is your biggest customer. It's a hopeless situation. We're selling Accurate Clock to a conglomerate who'll pay us for what's left of it."

"Let me see our contract with your bank," demanded Tom. Maria turned to a file cabinet to her left and in a moment pulled out a bound set of papers which she handed to Tom. He glared at Mortimer and sat down to look through it. Mortimer shrugged his shoulders and resumed his conversation with Maria.

Ten minutes later, Tom looked up. "This contract says you have to notify us in writing of any breach of contract," he said. "Then you have to give us 30 days to fix the problem before you can take action against us."

"Frank was willing to let your father go up to six months late, but he didn't write anything down," Mortimer said. "I, however, like to cover my bases, so I gave your father the notice on March 31, the day after he decided to miss his payment. It doesn't matter anyway. Today is day 12. There's nothing you can do in the next

18 days to fix this situation*."

"It's still my company," said Tom, "and I want you out of here now so I can work on turning it around."

"I heard you were fired from your company in New York," said Mortimer. "Are you trying to ruin your reputation even more? As we speak, your customers are looking for new suppliers they can count on. Your suppliers are cutting you off because they haven't been paid in months. You can't fix this company fast enough to reverse that. Only a conglomerate with deep pockets can provide instant confidence to your customers and suppliers and save what's left of this company."

Tom stared hard at Mortimer. "I want my 18 days," he said.

Mortimer gritted his teeth and curled his lips into a snarl. He whirled and threw the file he was holding against the wall. As the papers settled onto the floor, he leaned over the desk and stared at Tom. "That's a stupid decision," he said. He grabbed his jacket off the coat rack and stormed out the door.

Tom looked out the window. Maria watched him as they both listened to Mortimer's footsteps fade down the hallway. A car door slammed and Mortimer's Cadillac roared out of the parking lot.

---

* Conveniently, it is April 12 in our story so the 30 days ends on April 30.

"You sure you want to do this?" asked Maria.

"He was right, you know," Tom said. "I was fired from my job in New York. I only came here to sell this company and get back there. Now I've got nothing. I need to do this."

"Okay," she said quietly.

"Any idea where we should start?" asked Tom.

She leaned back in her chair and thought for a moment. "Actually, I met someone through another client who might be able to help us," she said. "His name is Roger Dominick. He has helped turn around several manufacturing firms as a consultant, but he told me he's looking for part ownership next time. Our management team should meet him and consider giving him part ownership in exchange for his help."

"I don't want to give up any ownership," said Tom.

"Unless we get help from someone like him," Maria said, "you'll lose all your ownership. We can't do this ourselves." She looked at Tom and waited.

Tom thought for a moment, then nodded. "Can he be here tomorrow morning?"

Value-Added Analysis

Tuesday, April 13   7:00 a.m.

All nine members of the management team were waiting in the conference room. "Cecil introduced me to you last week," Maria said to the group, "and you met Tom yesterday during his plant tour. To my right is Roger Dominick, the gentleman we told you about when Tom and I organized this meeting. I must also tell you that Tom and I dismissed the manufacturing manager yesterday. Tom will tell you why."

Tom stood. "You know we've been losing money for over a year. What you don't know is that four days ago West-Mart decided to stop buying clocks from us because our costs were too high, our quality didn't meet their standards, and our shipments were consistently late. They represented 47 percent of our revenues across all of our product lines."

"We lost West-Mart?" asked Kim Nguyen, production control manager. "We're dead."

"Not yet," said Roger. "We can still recover, but we must act quickly."

"We can't afford any false steps," said Maria, "and that is why I invited Roger to meet with us today. Tom and I discussed hiring him as president, making him a part owner, and putting him in charge of the effort to turn this company around. Before we make a final decision, we want him to spend this morning with us to describe his improvement strategy and give you a chance to grill him. We'll ask for your feedback, then make a final decision by the end of the day. Any questions?"

They all shook their heads.

"Roger, you're on."

*   *   *

"There are three areas we need to improve in this plant," Roger said. "Do you know what they are?"

"One area is quality," said Paul Shivers, head of quality control.

"Another is cost," said Maria.

13

"The third is delivery," said Kim.

"Right, right, and right," said Roger. "Those three key measures determine the profitability of any business. I remember them by using the acronym QCD."

"Fine," said Tom impatiently, "so what's your strategy?"

"Simple," said Roger dramatically. "Eliminate non-value-added process steps. By doing that we will improve quality, cost, and delivery. In a typical manufacturing company, value is being added to a product less than five percent of the time it is in the plant, and from what I've seen, your plant is no different*."

"So you're saying we're wasting time 95 percent of every day," said Roy Esterhaus, chief engineer.

"Yes," agreed Roger. "Not only 95 percent of your time, but 95 percent of the money you spend on the shop floor. And unfortunately, 95 percent is probably optimistic."

"But...," started Roy defensively.

"Hold on, Roy," Kim interrupted. "What do you mean by 'value-added,' Roger?"

---

* Value-added (VA) process steps are more visible than non-value-added (NVA) process steps, therefore traditional managers and engineers spend most of their time making the value-added steps more efficient. Given that the VA steps account for only 5 percent of the time, and the NVA steps account for 95 percent of the time, this behavior makes little sense.

"A process step is value-added if it causes a change in the physical state of the material, in accordance with customer specifications," said Roger.

"That's all we do," Roy said.

"We'll see," said Roger. "First we'll make a list of your process steps. Then we'll compare each step to our definition of value-added to see what kind of improvement potential exists here at Accurate Clock Company. That is called 'value-added analysis.'"

"Where do we start?" asked Evanson Lontubu, manager of marketing and sales.

"Start from the moment the vendor drops off the material," said Roger. "Estimate the amount of time spent on each step. Do as much as you can in an hour, then we'll look at the results."

The management team spent the next hour making a list of process steps and putting them in order. In an hour they agreed on the first 38 steps (see chart on next page).

"Good job," Roger said as he reviewed the list. "These are normal process steps for a manufacturing company. Let's start at the top. Is moving the material off the truck value-added?"

"Of course," said Lois Duke, the HR manager. "We wouldn't pay someone to do it if it wasn't."

15

#	Process Step	Duration
1	Move off truck	15 min
2	Wait in receiving	18 hours
3	Compare to purchase order	20 min
4	Wait in receiving	1 day
5	Move to quality control	15 min
6	Wait in quality control	2 days
7	Inspect	30 min
8	Wait in quality control	1 day
9	Move to raw material storage	15 min
10	Wait in raw material storage	28 days
11	Print work order	5 min
12	Attach work order	10 min
13	Move to marshalling area	15 min
14	Wait in marshalling area	2.5 days
15	Move to press #1	5 min
16	Wait at press #1	18 hours
17	Set up press #1	3 hrs
18	Load press	30 min
19	Stamp material	1 sec
20	Wait at stamping	1 day
21	Move to stores	10 min
22	Wait in stores	7 days
23	Print work order	5 min
24	Attach work order	10 min
25	Move to press #2	15 min
26	Wait at stamp #2	18 hours
27	Set up press #2	4 hrs
28	Stamp material	1 sec
29	Wait at stamp #2	18 hours
30	Move to stores	15 min
31	Wait in stores	7 days
32	Print work order	15 min
33	Attach work order	10 min
34	Move to deburr	15 min
35	Wait at deburr	18 hours
36	Deburr material	10 min
37	Wait at deburr	18 hours
38	Move to stores	15 min

"Is that how we determine if a process step is value-added?" asked Roger.

"Well, no," she admitted. "You had a different definition. I've got it in my notes. 'A process step is value-added if it causes a change in the physical state of the material, in accordance with customer specifications.'"

"Right," agreed Roger. "So is moving the material off the truck value-added?"

"Not according to your definition," she said.

"Why?"

"It doesn't cause a change in the physical state of the material," she said.

"Exactly," said Roger. "My definition of value-added is tough, but it will force us to look critically at our process and uncover improvement opportunities. The next process step is 'wait in receiving'. Is that value-added?"

"Not according to your definition," said Sam Mordecai, purchasing manager, "but you can't expect my receiving people to drop everything they're doing every time a shipment comes in."

"Does that mean having the material sit there is value-added?" asked Roger.

"No," admitted Sam, "but you make it sound like we're letting things sit for no reason."

"That's the way your current process works," said Roger. "Could your process be changed so that the material doesn't have to wait?"

"I don't see how," said Sam.

"We'll get to that," said Roger, "but for now do you agree that having material sitting around is non-value-added?"

"Yes," said Sam.

"Okay, how about comparing the incoming material to the purchase order?" Roger asked.

"If we don't check it against the P.O., we might mistakenly accept the wrong material," said Evanson.

"Does that make this step value-added?" asked Roger.

"No," said Evanson, "because we aren't causing a change in the physical state of the material."

"Right," said Roger. "Then the material waits in receiving again."

"Non-value-added," said Sam quickly.

"Then we move it to quality control," said Roger. "Value-added?"

"If we don't move it," said Roy, "how on earth is it going to get there?"

"Just because we have to move it in our current process doesn't make it value-added," said Kim. "Besides, maybe we could move quality and receiving next to each other, right Roger?"

Roger smiled. "You're getting it now. Next, the material waits in quality control..."

"Non-value-added!" they all said at once.

"Very good," said Roger. "How about 'inspect'?"

"Well, we've got to inspect the material," said Paul. "We don't want bad material from the vendor going into our products."

"That doesn't make it value-added," said Kim. "Inspection doesn't change the physical state of the material. Besides, our vendors should make sure it's right before they send it to us. What does everyone else think?"

They all nodded in agreement.

"Steps 8, 9, and 10 are wait, move, and wait," said Roger. "Non-value-added?" Everyone nodded. "What about printing the work order?"

"That causes a change in the physical state of the material," said Roy. "The ink goes on the paper, and it's stuck on there for good."

"But it doesn't physically change the product," said Sam.

"That's right," said Roger. "A process step is not value-added unless it changes the physical state of the material going into the product itself."

"So the next step is non-value-added also," Evanson

said. "Attaching the work order doesn't change the material."

"Right," agreed Roger. "Steps 13, 14, 15, and 16 are move, wait, move, and wait, so we know they're non-value-added. How about number 17, setting up press number one?"

"If we don't set up the machine, we can't stamp the metal," said Roy.

"So setting up the press is value-added?" asked Roger.

"I know what you're going to say," said Roy. "You are going to tell us that setting up the press doesn't change the material, so it's not value-added. You've been telling us that everything we do is a waste of time, like we're a bunch of idiots. So far we've done 17 steps, and you say that not one of them adds any value."

"You noticed," said Roger.

"I'd like to see you go down on the shop floor and tell our guys to their faces that setting up a machine is non-value-added," Roy said. "See what they tell you. They spend more time setting up machines than running parts."

"Every step you do on the shop floor is necessary given your current process," Roger said. "That doesn't mean the steps add value. I bet the guys on the shop floor are more frustrated than anyone in this room because they spend so much time setting up the machines."

"That's true," admitted Roy.

"Human beings get a whole lot more satisfaction out of transforming material into a product than they do out of generating paperwork, moving parts, and inspecting material that should have been made right in the first place," said Roger. "Not only is non-value-added bad for business, it's bad for morale. The waste is all around us, but we can't see it anymore. We've learned to live with it. Looking at our processes using our new definition of value-added will wake up our minds and show us where to improve."

Roger looked around the table. The management team was quiet.

"Makes sense," said Maria.

"The next step is loading the material onto the press," said Roger.

"Non-value-added," said Lois, "but necessary in our current process."

"Right," said Roger. "How about stamping the material?"

They all looked at each other.

"Value-added?" said Sam.

"Maybe," said Roger, smiling.

Roy looked confused.

"It causes a change in the physical state of the material," said Kim, "but the other part of the definition of value-added requires it to be 'in accordance with customer specifications.'"

"Okay," said Roy, looking relieved, "as long as we do it right, it's value-added."

"Right," said Roger.

"We finally added some value," said Roy.

"Hooray!" they all yelled.

"You can analyze the rest of the process steps without my help," said Roger. "Do the math and tell me what percentage of time spent in the first 38 steps is value-added. Then tell me what the biggest source of non-value-added time is."

The management team huddled together over the table for a few minutes.

"We've got it," said Maria. "The total elapsed time is 54.5 days. The total value-added time is 10 minutes 3 seconds. The percentage of time we are adding value is .01 percent. That's much lower than the 5 percent you said in the beginning."

"That's not unusual," said Roger. "What is the biggest source of non-value-added time?"

"Fourteen out of 38 steps are waiting," reported Maria. "Waiting time accounts for 54 days, or 98 percent of the non-value-added time."

"Well, great," said Roy. "What a wonderful discovery. Too bad we can't get rid of it. Our MRP* system makes us order parts in advance, so we'll always have them when we need them."

"Do you?" asked Roger.

"Do I what?" said Roy.

"Always have the parts when you need them?" asked Roger.

"Well, no," responded Roy. "We're always missing a few parts, which is why we never ship on time."

"Who told the MRP system how far in advance to make the parts?" asked Roger.

"I did," said Kim.

"So you told it what to do, and now it tells you what to do?" asked Roger.

---

* MRP stands for Materials Requirements Planning. It is software that uses demand forecasts, bills of material and lead time estimates to determine when to order parts from suppliers, release parts to the floor, and build subassemblies in order to deliver products on time. MRP does not work very well for shop floor management because it assumes a predictable environment. It has been said that no battle plan survives contact with the enemy. The enemy of MRP is reality, and that is why traditional manufacturers do so much expediting. Traditional manufacturers use MRP to deal with problems by ordering extra parts and lengthening the estimated lead times. This gives them "more room for error". Lean manufacturing (as we shall see) actually allows less room for error, forcing you to deal with the underlying causes of problems immediately so that they are solved permanently. Later versions of MRP are MRP II (Manufacturing Resources Planning), and ERP (Enterprise Resources Planning).

"That's the way our system works," said Roy. "What do you want us to do, wait until the last minute to make the parts we need?"

"Yes," said Roger.

"What?" said the entire management team simultaneously.

"You heard me," said Roger, chuckling. "But we'll get to that later. Let's keep talking about eliminating non-value-added process steps for now."

"So we've got a lot of steps to eliminate," said Maria. "I have another question. Earlier you said that by eliminating process steps we will improve quality, cost and delivery. Explain that."

Roger turned to the board behind him and drew this diagram:

□---->□---->□---->□---->□---->□---->□---->□

"Pretend this is a map of the process we use to produce our product," Roger said. "The boxes are the process steps, and the arrows show the order in which they are performed. This process has eight steps. Let's pretend we found a way to eliminate half of the process steps. Then the process would look like this:"

□---->□---->□---->□

"What happened to the quality of our product?" asked Roger.

"You want us to say it gets better," said Roy, "but I don't know how it could."

"I know," said Sam. "Every process step is an opportunity to make a mistake. The more steps, the more mistakes."

"What about the non-value-added steps?" asked Roger. "Is it possible to create a defect even when you are not adding value?"

"Sure," Sam said. "You could copy a number wrong, and the next thing you know, the wrong shipment could be delivered to the wrong customer. Or you could crash the forklift while moving parts around."

"You've got it – so much for quality," said Roger. "So what happened to the cost of manufacturing our product when we eliminated half the process steps?"

"It would obviously be less," said Maria.

"Why?" asked Roger.

"Because every step costs something," said Maria. "We have to pay someone or use chemicals or heat or something for every process step."

"Even the non-value-added steps?" asked Roger.

"Sure," said Evanson. "When we move material we pay labor, use electricity, and add mileage to the forklifts."

25

"What about the waiting steps?" asked Lois. "How do they cost us money?"

"Easy," said Maria. "Material sitting around is inventory. First, it takes up space we could use for something else or rent out. Second, there is money tied up in the material we could invest. Finally, we run the risk of the material getting damaged or becoming obsolete. I bet one-third of this building is taken up by motionless inventory."

"So every step we remove saves us money," said Roger. "Even the non-value-added steps. What happened to our delivery time when we removed half the steps?"

"It obviously would decrease because fewer steps take less time," said Lois.

"Right," said Roger. "Now you know that our manufacturing process contains many non-value-added steps, and that by eliminating them, we will improve our quality, cost and delivery."

"Makes sense to me," said Tom.

"One more point while we're on this subject," said Roger. "We are not going to forget about quality or cost, but delivery improvement, or lead time, will be our first priority, and the goal we stress to our people."

"Why shouldn't we have quality or cost be our most visible measure?" asked Tom.

"If you tell traditional manufacturing people that you want to improve quality by 50 percent, what would they want to do?" asked Roger.

"Add more inspections," said Paul, "which is what I've been saying for the past six months."

"That wouldn't work," said Sam. "That's more non-value-added work. Inspections take time, cost money, and believe it or not, you can create a defect during inspection."

"Good. What if you told a group of traditional manufacturing people that you want to cut costs by 50 percent?" asked Roger.

"They'd lay off a bunch of people and stop making any investments in the future," said Lois.

"That's right," said Roger, "but let me ask you something. What happens if we ask our people to help us improve lead time by 50 percent?"

"We can't do it by working faster," said Tom.

"The only way you can do it is by eliminating process steps," said Roger. "There is no other way to do it. And when we eliminate steps, we improve what?"

"Quality, cost, and delivery!" yelled Kim, Sam and Lois together.

"Right," said Roger, grinning ear to ear. "Any questions?"

"Yes," said Kim. "This will work great if we can

figure out how to eliminate those non-value-added steps. Do you have any tools to do that?"

"As a matter of fact, I do," said Roger. "Let's take a break and then we'll talk about them."

CHAPTER SUMMARY

Three measures:

1. Quality (defects, rework, returns)
2. Cost (materials, productivity, overhead)
3. Delivery (lead time)

Strategy to improve QCD:

Eliminate non-value-added process steps.

Value-Added definition:

- *A process step is value-added if it causes a change in the physical state of the material, in accordance with customer specifications.*

- This tough definition forces people to look at their processes critically and uncover improvement opportunities.

- In a typical company, value is being added to a product for less than 5 percent of the time the material is in the plant.

Process map

□--->□--->□--->□--->□--->□--->□--->□

If you eliminate process steps

□---->□---->□---->□

You get…

Better quality
Every step is an opportunity for a mistake

Better cost
Every step costs money

Better delivery
Every step takes time

Lean Ingredients

"The best way to eliminate non-value-added steps is to implement a set of techniques known as *lean manufacturing,**" Roger announced.

"Sounds like another couple of buzzwords to me," said Roy. "You want to come in here and experiment on us with the latest trend that's supposed to save the world."

Roger laughed. "I know how you feel," he said.

---

* An exception: if you are managing a high-mix, low-volume plant, you must implement 5S, standard work, setup reduction, and preventive maintenance. However, just-in-time production, continuous one-piece flow, work cells, and kanban can actually *decrease* efficiency in a true high-mix, low volume environment because these techniques require some degree of repetitiveness.

"However, I'd categorize the phrase 'lean manufacturing' as a brand name rather than a buzzword. Although the phrase 'lean manufacturing' is new, the underlying techniques are anything but the latest trend. They were first implemented by Taiichi Ohno at Toyota in the 1950s."

"What did you mean when you said 'lean manufacturing' is a brand name?" asked Emerson.

"Human beings get bored with any idea that has been around for a while," explained Roger, "and we tend to think that new ideas work better than old ideas. Therefore, in order to impress potential clients, consultants invented new brand names to sell the old techniques behind lean manufacturing.

"Lean manufacturing is now the hottest brand name, Just-In-Time (or J.I.T.) is less hot, and the Toyota Production System (TPS)* is making a comeback," said Roger. "Other brand names include cellular manufacturing and modular manufacturing. They are all the same, so I don't want you to be confused by them."

"But how do you know they are all the same?" asked Paul.

---

* TPS consultants have strict definitions and rules of implementation. This book is not an explanation of TPS. It is a common-sense explanation of the underlying lean manufacturing ingredients and how you can implement them quickly to improve your operation.

"Because they all require the same ingredients to make them work," said Roger, "and if two things contain the same ingredients, they are the same thing."

"And those ingredients are...?" prompted Lois.

"Glad you asked," said Roger. He turned to the board and wrote this list:

- Just-in-time production
- Continuous one-piece flow
- Work cells
- Setup reduction
- Preventive maintenance
- Kanban
- Workplace organization and cleanliness
- Standard work
- Teams of employees who think and take initiative

"I've heard of most of those," said Kim, "but I don't know exactly what they mean."

"I'll explain each ingredient in its ideal state," said Roger. "Our job will be to make continual improvements and work toward those ideals. We will be forced to make some compromises, but as we get closer to the ideal, our jobs will get more secure, and our profits will increase.

"Just-in-time production means we will make what the customer wants, when the customer asks for it. No more making products ahead of time because we think the customer might want them some time in the future."

"OK, stop right there," said Roy. "Right away I see three things wrong with that."

"Let's take your objections one at a time," said Roger. "What's the first one?"

"First, if we don't have the product they want in stock, they will have to wait while we make it from scratch," Roy explained. "We told you the first 38 steps of our process alone take 54.5 days. Our whole process takes six months. No customer is going to wait that long."

"The second lean ingredient will take care of that," Roger said. "Continuous one-piece flow means once material enters the building, it never goes into stock. It never stops moving until it is a finished product, at which time it is shipped to the customer. This means no more batch processing. Once an operation is completed on a piece of material, that material moves immediately to the next operation. This will reduce our total lead time dramatically, because we know from our value-added analysis that 98 percent of our lead time is waiting."

Roy listened patiently. "Nice theory," he said, "but if we moved each part to the next machine every time we finished processing it, all we'd be doing all day is moving parts around. Sometimes the next machine is on the other side of the plant."

"You answered your own objection," said Sam. "We have to move the machines closer together."

"Exactly," said Roger. "A work cell is a group of all the machines or work stations necessary to make an entire product, or a large component of an entire product. They are placed as closely together as possible so that parts can be passed to the next operation as soon as the previous operation is completed."

"I can tell you right now why that isn't going to work," said Roy. "We don't have enough equipment to dedicate to individual products. We probably have enough for two or three dedicated cells, but we have forty different products."

"There are two strategies to deal with that objection," said Roger. "The first is to sell the new, high-tech machines we have and replace them with several smaller, older, less expensive machines that still can do a quality job. The older machines are not as fast, but they don't need to be. They will only have to serve one product line."

Sam laughed.

"What's on your mind, Sam?" asked Evanson.

"As we've brought in new machines, we've been

putting all our old machines in the back corner of the warehouse," Sam said. "The other day, we were talking about selling them for scrap. I'm glad we didn't."

"Me too," said Roger. "If we must share a machine across more than one product line, then a compromise strategy is setup time reduction. That way we can make parts in small batches for each product line as we need them."

"So what you're saying is these work cells are going to be dedicated to making a single product line...," Roy began.

"Or a product family," Roger interjected.

"Right," agreed Roy, "and for each product line there will be only one machine available for a given operation."

"Right," said Roger.

"Hah! So what if one of the machines in the work cell breaks down?" asked Roy. "The way we're set up now, we share machines across all product lines. If one machine isn't working, we move the work to another machine. We can't do that if we do what you're saying."

"I have to admit, you're right," said Roger. "That's why the next ingredient, preventive maintenance, is critical. We can't have machines breaking down in a lean manufacturing environment. Preventive maintenance not only will eliminate unexpected breakdowns, but also will increase the life of the machine. Any more objections, Roy?"

"Not right now, but I'll think of some," Roy promised.

"Anyone else?" asked Roger.

"What exactly is 'kanban'?" asked Lois.

"Like setup reduction, kanban is a compromise," explained Roger. You only use it if you can't figure out how to make an entire product within one dedicated work cell. In reality, kanban is nothing more than controlled work-in-process (WIP) inventory stored between two work cells. When the inventory gets low, the supplier cell makes more parts for the customer cell."

Lois looked confused. "Can you give us an example?"

"Sure," said Roger. "Let's say that a work cell in a motorcycle company makes gas tanks. They have two special padded boxes that I will call 'kanban containers'. They each hold five gas tanks. The work cell makes enough gas tanks to fill both kanban containers and then leaves them next to the person who installs the tanks on the assembly line. Then the supplier work cell stops making gas tanks, but they keep an eye on those kanban containers. As soon as one of the containers is empty, they grab it, make a batch of five more tanks, and deliver that full kanban container back to the assembly line.

Meanwhile, the person on the assembly line is using the gas tanks from the second container. When the second container is empty, it goes back to the work cell for more tanks while the person at the assembly line uses the tanks from the first container. It goes on like that forever."

"I get it," said Lois. "It seems like what you want to do is eventually reduce the number of tanks down to one in each kanban container until you've got it so that each time the person at the assembly line uses a tank, the work cell delivers the next one. Is that right, Roger?"

"Absolutely," Roger said. "There are many different ways to use kanban to connect cells together, but they all involve controlled WIP and some way of signaling the need for more parts*."

"Seems to me that kanban requires a smooth and predictable flow of orders or else you'd have too many or not enough parts," said Kim.

"That's right," said Roger. "Your internal operations and external suppliers better be working smoothly as

---

* The above example is called "two-bin" kanban, in which the empty container itself serves as the signal to make more parts. An alternative method is to attach a "kanban card" to the full container when it is delivered to the assembly line. As soon as the first part is taken out of the container, the card is sent back to the supplier cell so they can start working on the next batch of tanks. The method can result in less WIP because there will never be two full containers waiting at the assembly line.

well. If you have quality problems in the supplier cells, or are frequently running out of components, your smooth flow will be interrupted and kanban won't work."

"The next ingredient on the list is workplace organization and cleanliness," said Roy. "I won't argue with that."

"That is also known as '5S,'" said Roger, "because the Japanese have five words all beginning with 'S' that describe their strategy for attaining workplace organization and cleanliness. Most American consultants translate the words and the underlying meanings into English as faithfully as possible. I do the same for the first three S's, but I've changed the words and meanings of the fourth and fifth S's*."

"The first S is Sort, which means get rid of anything we don't need in the workplace. The second S is Straighten, which means to establish a place for everything and keep everything in its place. The third S is Scrub, which means get everything very clean and keep it that way. The fourth S is Schedule, which means to establish a daily routine in each work cell to maintain and improve on the first three S's. The fifth S is Score, which

---

* The five Japanese S's are spelled in English as *Seiri, Seiton, Seiso, Seiketsu*, and *Shitsuke*. The last two S's are most frequently translated as Standardized Cleanup and Discipline, but the typical Japanese explanations are not useful.

39

means to regularly measure how well each cell is doing at maintaining and improving the first three S's."

"Standard work means documenting the best way we know to do every task in the plant, and making sure each task is done that way every time," explained Roger. "If we don't establish and use standard work, all our improvements will disappear because everyone will do their work differently."

"But what if we figure out a better way to do a task after we've documented the standard work?" asked Maria.

"We change the standard work for that task," said Kim. "That's the whole idea."

"Right," agreed Roger. "Standard work is the highest quality, lowest cost, and fastest way to do work. If we can improve on it, we will. Unless your supervisor has given approval to temporarily experiment with a different method, you'd better be doing the job according to standard work."

"It seems to me that 5S and standard work are pretty basic ideas," said Maria.

"Yes they are, yet they are not established here at Accurate Clock Company," said Roger. "You should not feel bad though, because very few companies have done so. Many managers get caught up in the 'latest and greatest', and they forget to pay attention to the basics."

"But if you can't maintain the basics," said Maria, "how can you expect to implement the more complicated techniques?"

"You can't," said Roger.

"What's this 'team' nonsense on your list?" asked Roy. "We tried teams here before, and they were a miserable failure."

"How did you go about trying to implement them?" asked Roger.

"We had a local community college come in and train all our hourly employees in communication skills, conflict management, and problem-solving," explained Lois. "Then we went on a trip to the mountains for some adventure-based learning. We fell backward into each other's arms and jumped off poles and crossed narrow bridges with ropes tied to us. Then we started calling all our departments 'teams' and all our supervisors 'coaches.'"

"And that worked?"

"Well, no," admitted Lois. "Nothing changed. In fact, many employees said it was a waste of money."

"I'm not surprised by your experience," said Roger. "In fact, studies show that 87 percent of all efforts to implement teams are unsuccessful. Nonetheless, teams can be a powerful competitive weapon. Teams will be necessary for us to successfully implement lean manufacturing, because lean manufacturing requires hourly employees who can solve problems without help from management every few minutes."

"So how will we succeed this time with teams?" asked Lois.

"Do you believe that the process used to produce a result determines the quality of that result?" asked Roger.

"Yes," said Lois.

"The problem with your effort to implement teams was the process you used," explained Roger. "I call your approach 'traditional teambuilding', and it never works. We will do it differently this time. We will start by teaching our supervisors how to be coaches."

"What do you mean by 'coach?'" asked Lois.

"Most supervisors are stuck in the 'babysitting' mode," said Roger. "They treat their people like children. Coaches treat employees like adults. Coaches refuse to solve problems and make decisions on the shop floor, and instead expect their people to solve the problems and make the decisions. Until your supervisors become coaches, teams will never work because your supervisors won't let them work.*"

---

* For more about this non-traditional approach to implementing teams, see the appendix for an excerpt from *The Journey To Teams*, also by Michael D. Regan.

"Makes sense," said Lois.

"Any more questions about lean manufacturing?" asked Roger.

"I've got a big one," said Lois. "As soon as you said the word 'lean' it sounded to me like you meant 'layoffs', as in doing work with as few people as possible. Does it mean that?"

"Great question," he said. "Let's take a quick break and then talk it over."

CHAPTER SUMMARY

The best way to eliminate non-value-added steps is to implement lean manufacturing.

Lean manufacturing is synonymous with:
- Just-In-Time
- Toyota Production System (TPS)
- Cellular Manufacturing
- Modular Manufacturing

Lean ingredients:
- Just-in-time production
- Continuous one-piece flow
- Work cells
- Setup reduction
- Preventive maintenance
- Kanban
- Workplace organization and cleanliness
- Standard work
- Teams of employees who think and take initiative

The Layoff Decision

"Once we start to implement lean manufacturing, we must promise not to lay off anyone as a result of productivity improvements," said Roger.

"No way," said Tom. "I thought you were going to tell us how to get profitable again. The only way to make a manufacturing company more profitable is to cut as many people as possible."

"That is one of the dumbest statements I've ever heard," said Roger bluntly.

"I have to admit," said Sam, rising to Tom's defense, "I always thought cutting people was management's goal. We justify all our new machines based on how many people we'll be able to replace."

"That's got to stop if you want to be survive," said

45

Roger.

"Explain yourself," said Roy.

"When we implement lean manufacturing, a big part of our cost savings will be from productivity improvements, about 20 percent each year," said Roger. "Your employees aren't dumb. They know if productivity goes up and volume stays the same, the number of employees must decrease. We need our employees to help us implement lean manufacturing, and they're not going to help if success means layoffs. There is only one solution to this problem."

"Growth," said Evanson.

"Growth?" asked Tom.

"If we improve productivity, the only way to maintain or increase the number of employees is to bring in more work," explained Evanson. "That means I have to sell more. I can sell more if our quality, cost, and delivery get better. That's what lean manufacturing will do for us."

"Not only do I want you to sell more," said Roger, "I want you to visit the shop floor regularly to explain how your sales efforts are progressing."

"No problem," agreed Evanson.

"I do have a problem," said Tom. "We just lost 47 percent of our business. We aren't going to be able to go out and replace that business tomorrow, especially since we don't have any quality, cost and delivery improvements yet. Are we going to pay 47 percent of our people for doing nothing?"

"Tom," said Evanson, "Roger said 'no layoffs after we start lean manufacturing.' We haven't started yet."

"Exactly," said Roger. "If layoffs are necessary, we must do it now and get it over with. I recommend that we lay off 47 percent of the workforce immediately, and the pain has to be shared equally between the salaried and hourly ranks."

"Wow," said Roy, "that's 94 people. This is serious."

"I don't want to do it," said Roger, "but unless the bank sees us taking quick action, there won't be a job here for anyone. They'll sell us to a big conglomerate who'll take our brand names, products, and customer list and close this place down."

"Of our 200 employees," said Lois, "55 are temporaries, so we can let them go today. We also have 32 people on warning who should have been fired a long time ago. That means we'll have to lay off seven good employees with the least amount of seniority."

"They will be the first to be brought back once we start hiring again," said Maria. "We're on the brink of bankruptcy. Either we let a few people go now to give us a chance to recover, or we let everyone go when the entire business fails."

"Let's get it over with and start moving forward," said Lois.

"It's 11:00 a.m., and we all have work to do this afternoon," said Maria. "Roger, can you give us a few minutes to discuss our decision about offering you the job as president of Accurate Clock Company? You can wait in Tom's office."

"Sure," said Roger. "If you decide to hire me, bring Evanson with you. We'll need to talk about sales." He gathered his notes and left the room.

Maria looked around the room. "This is a decision that Tom and I must make," she said, "but I want to hear your opinions first."

"We don't want anything to do with him," said Paul. "His ideas will sink us. We'll lose all our efficiencies if we don't make parts in batches. I can tell you our problem – too many lazy people who don't care."

"But Roger says people only appear to be that way because of how management treats them," said Lois. "We don't give them a chance to think and solve problems. We always do it for them."

"I agree with Paul," said Roy. "Roger is wrong for this company. He obviously doesn't understand our process. Our setup times are as low as they can be, and we don't need to reduce inventory, we need to build more. We run out of parts all the time now due to rework or bad scheduling."

"I totally disagree," said Evanson. "He is perfect for the job. Better quality, faster lead time, and lower costs will help me sell a lot of product."

"That's just it," said Paul. "His ideas won't get us there."

"Yes they will," said Kim. "You and Roy are stuck on your old ways of thinking. All you two did today was ask stupid questions and you didn't even listen to his answers."

"Now just a minute," said Paul. "He was talking nonsense. Especially all that 5S stuff. As if a little housekeeping is going to solve all our problems."

"5S was the only thing I agreed with!" said Roy.

"You've got to be kidding," said Paul. "The only reasonable thing he talked about was standard work."

"Yeah right," said Roy sarcastically. "That would be a bureaucratic nightmare."

"We've heard enough," said Maria. "Let's see a show of hands for everyone who is against hiring Roger."

Paul and Roy raised their hands while they glared at each other.

"Now who wants to hire him?"

Lois, Evanson, Sam and Kim raised their hands. So did Tom and Maria.

"Great," said Paul sarcastically.

CHAPTER SUMMARY

No Layoff Policy

- **The Promise**: No layoffs due to productivity improvements.
- Higher productivity and no layoffs require **growth**.
- Better QCD will help sales.
- If layoffs are necessary, do them before starting lean manufacturing activities.

Staying Alive

"Roger, we'd like to offer you the job," said Maria. "We'll pay you a minimal salary, but you can earn up to 30 percent ownership in the company over two years."

"I accept," he said. "Now let me ask you a few questions. How long can we meet payroll with the cash we have on hand?"

"Six weeks with the layoffs we're doing this afternoon," said Maria.

"Payroll must be our first priority," said Roger. "We need to treat our remaining people like gold. If we lose them, we're dead for sure."

"What about making debt payments to get the bank

off our backs?" asked Tom.

"Can't do it," said Maria. "One loan payment is equal to a week and a half of payroll. If we make two loan payments, one for last month and one for this month, we'll only have three weeks for payroll. That's not long enough to make progress."

"And six weeks is?" asked Tom.

"You'd be surprised," said Roger.

"How do we get Mortimer Stern off our backs?" asked Tom.

"We need to show the bank that we're their best chance to recover their money," said Roger. "At best, they'll only get 50 cents on the dollar in a sale to a conglomerate. We've got to show that we are taking immediate steps to stop the bleeding. The layoffs were the first step. More important, we need new customers. Evanson, have we got any hot prospects?"

"Lots of retailers want clocks," said Evanson, "but our prices are too high because our costs are too high."

"What if we dropped our price by 10 percent today?" asked Roger.

"Then I could get some contracts signed."

"By when?"

"Within the next couple of weeks."

"Go do it."

"But what good would that do?" asked Tom. "We're

barely breaking even on most of our product lines now. We'll be losing more money."

"No we won't," said Roger. "I've seen the shop floor, and I can tell you that our first improvement efforts on one product line will cut our costs by 20 percent, maybe more."

"Impossible," said Maria.

"Show up at 7:00 a.m. tomorrow and we'll continue the education," said Roger. "This afternoon Tom and I will head downtown and meet with Frank Fowler at Indiana First Trust. We can use his loyalty to Cecil to soften him up. Either way, we need to tell him what we're doing."

Lean Implementation Strategy

Wednesday, April 14   7:00 a.m.   Conference Room

"Roger accepted our offer to become our new president," announced Tom. "And we are going to implement lean manufacturing as our turnaround strategy. Roger, where do we start?"

"Let's discuss the overall strategy we will use to implement lean manufacturing," said Roger.

"First, we will teach everyone 5S and start implementing it everywhere in the plant. No one will argue with organization and cleanliness, and we'll get everyone involved in producing immediate, visible results."

"Second, and we'll do this in parallel with the first

step, we'll choose a product for which we will implement all of the lean ingredients from final assembly back to receiving. This is called a model line, and it will serve as an example or pattern for the rest of our products. How do you think we ought to choose our first product?"

"It should be one that has many problems, high volume, and growth potential," said Evanson, "so that we get big benefits from the improvements we make."

"It also should be one of our less complex products and processes, so that our first implementation is not overwhelming," added Sam.

"I like the way you guys think," said Roger. "In our case, Evanson is working hard to bring in new orders to replace the revenue we lost from West-Mart. We are offering a 10 percent price break, so we'll need to concentrate on whatever product he sells.

"Third," said Roger, "we will implement work cells, continuous one-piece flow, and standard work in final assembly."

"Why start with final assembly?" Paul asked.

"Because smooth and consistent production in final assembly will make the work cells in component fabrication easier to manage," Roger said.

"Fourth, we will build work cells further upstream in component fabrication and use kanban to connect them to final assembly. We will progressively reduce setup time and batch sizes in these cells and implement preventive maintenance.

"Fifth, we will apply everything we learned on our first model product line to all of our other products*."

"Roger," Lois said, "all that makes lots of sense, but it's more change than we've ever attempted around here. We'll be out of business before we get one-tenth of it completed for one product. It took us three meetings last month to decide what kind of coffee to serve in the cafeteria. With all the fires we fight around here, it's hard enough to get all the right people in the same room."

Roger nodded and smiled. "Let me tell you about kaizen events, Lois."

---

* This book takes place at a fictional company that makes clocks. Lean manufacturing will probably not apply to your company in exactly the same way. Success with lean manufacturing will depend on your ability to keep an open mind and be flexible and creative about how to apply the lean ingredients. Doyle Wilson of Doyle Wilson Homes in Austin, Texas was the first to figure out how to apply lean manufacturing to his industry, and as a result he is wiping out his competition. If these ideas can be applied to home building, they can probably be applied to your industry.

CHAPTER SUMMARY

Lean Implementation Strategy

1. Teach everyone 5S and start implementing it everywhere in the plant.
2. Choose a product to become the model line. This will become the pattern for other product lines.
3. Implement work cells, continuous one-piece flow, and standard work in final assembly.
4. Build work cells in component fabrication and use kanban to connect them to final assembly.
5. Repeat steps 3 and 4 for all product lines.

Kaizen Events

"What process have you been using to implement improvements at Accurate Clock Company?" asked Roger.

"We schedule one-hour meetings once a week for a group of people to discuss ideas and make decisions," said Paul. "We call them 'quality circles.'"

"And that works?" asked Roger.

"No," said Paul. "It's impossible to get everyone to show up for every meeting, no one is ever on time, and soon they stop coming at all because they're bored and frustrated with the lack of progress."

"The worst part is that people forget what they discussed at previous meetings and we end up making the same decisions over and over again," added Lois. "In some meetings we accomplish absolutely nothing."

"We will still need to have a few short meetings," said Roger, "but most of our lean manufacturing implementation will be done using kaizen events.

"In a kaizen event we bring a cross-functional group of five to 15 people* together for a full week to completely implement several lean manufacturing ingredients for a certain process. For example, we will be able to implement work cells, continuous one-piece flow, and standard work in final assembly for one product in one week. The keys to a good kaizen event are good preparation before the event and good leadership during the event.

"A kaizen event utilizes each participant for 40 hours all at once as opposed to one hour a week for 40 weeks. This makes better use of the participants' time, and produces an immediate and sizable result."

"What kind of results are you talking about?" asked Tom.

"Glad you asked," said Roger, as he passed out copied sheets of paper to everyone in the room. "Here are the results from four different kaizen events I helped

---

* A group of eight or more people is too large to work effectively as a team. In a kaizen event requiring more than seven people, most of the work during the week is done in sub-teams of two to six people.

lead as a consultant*:"

Event A: Plastics Assembly
- 102 percent productivity improvement
- $1.2 million continuing annual savings

Event B: Sheet Metal Fabrication
- 73 percent reduction in WIP
- 49 percent reduction in floor space
- 85 percent reduction in setup time
- Lead time reduced from 23 to 10 days

Event C: Recyclables Processing
- 31 percent productivity improvement
- $300,000 continuing annual savings

Event D: Administrative Paperwork
- Lead time reduced from three days to 24 hours
- Exceptions reduced 50 percent

"Those kinds of results would certainly make my

---

* Actual results from four kaizen events in the automotive, aerospace, and textile industries for which Everest Consulting Group provided consulting.

marketing job easier," said Evanson.

"And they'd make our financial statements look much better," said Maria.

Tom looked at the numbers and smiled.

"What does the agenda for a typical week-long kaizen event look like?" asked Maria.

"On Monday morning we do lean manufacturing training and brainstorming. We spend Monday afternoon prioritizing improvement ideas and planning approaches. On Tuesday and Wednesday we experiment with ideas and implement those we like. On Thursday we make sure to finish everything we are working on. On Friday morning we clean, then give tours of the area to people from other departments to spread the ideas and the excitement. Then we have a celebration lunch."

"When is our first event?" asked Tom.

"As soon as Evanson tells us what product to focus on, we can start planning," said Roger. In the meantime, I am going to set up a training session for all employees, and then start implementing 5S plant-wide."

"There's no time to waste," said Tom. "Let's go have lunch, then get to work."

CHAPTER SUMMARY

Traditional Improvement Technique
Weekly, one-hour meetings for a group of people to discuss ideas and make decisions.

Kaizen Event
Cross-functional group of five to 15 people makes a drastic improvement in one process in one week.

Kaizen Agenda
- *Monday morning*: Lean manufacturing training and brainstorming.
- *Monday afternoon*: Prioritize improvement ideas and plan approaches.
- *Tuesday and Wednesday*: Experiment, test and implement.
- *Thursday*: Finish all projects.
- *Friday*: Complete 5S, give tours of the area to spread ideas and excitement. Celebration lunch.

Inspiration for the Shop Floor

Thursday, April 15   3:00 p.m.   Company-wide Training Session

"For years now, our customers have been asking us to cut our prices, improve our quality, and reduce our delivery times," said Roger to the 106 employees gathered in front of him. "We failed to meet their expectations, and now we've lost West-Mart and 47 percent of our volume. As you know, we were left with a painful choice: either close the plant, or lay off 47 percent of our coworkers and friends. We did the layoff.

"The manufacturing methods we are using today made us successful for the last 80 years, but now we need new manufacturing methods," Roger continued. "I

63

am here to lead that change, and I will introduce those new manufacturing methods to you in a moment. I don't expect you to immediately understand everything I am about to explain to you. I do expect you to listen carefully and do your best to learn as quickly as you can. These manufacturing methods have worked for me again and again, and they will work for us here. At first, you may not agree with everything you hear. I am not asking for your agreement today, but I will be asking for your commitment to help us implement these new methods.

"We will not get better by working harder. I assume everyone is already working as hard as they can. The techniques I am about to explain to you are different ways to work, and I promise you they are smarter ways to work. So keep an open mind, and listen closely."

For the next two hours Roger told them about value-added analysis, the lean manufacturing ingredients, and the overall lean implementation strategy.

"Our number one goal is to cut delivery, or lead time, by 90 percent for all product lines within two years," explained Roger. "We will not accomplish this goal by working faster, but by using the lean ingredients to eliminate non-value-added process steps, especially unnecessary waiting and moving. As we eliminate non-value-added process steps, we will improve quality in the next year from 70,000 defects per million clocks to less than 10,000 defects per million clocks, and to less than 5,000 defects per million clocks in two years. Elimination of non-value-added work will also enable us to meet our productivity improvement goals of 20 percent per year for the next three years.

"Productivity gains normally mean layoffs, but here they will mean growth," said Roger. "As long as I am here, I commit to never having a layoff at Accurate Clock Company due to productivity improvements. In turn, I need your commitment to me and to each other. I need you to be open-minded and try your best to understand and implement the lean manufacturing ingredients. I need you to tell us how we can reach our goals, not why we can't. And I need you to maintain a positive attitude at all times.

"I have a large piece of paper in my office with my commitment and signature on it. Below my name, I have written the commitment I am asking of you. If you are willing to make this commitment, I want you to come to my office before the end of this week to shake my hand and sign your name. We can do this with fewer of you, but I want to do this with all of you. We need your experience, and we need your brains. If you have not shaken my hand and signed your name by the end of the week, I will assume you have resigned and will send you one month's severance pay.

"Starting tomorrow, I personally will be working with each area on the shop floor to implement workplace organization and cleanliness, also known as 5S. That will give each of you an opportunity to get involved immediately in making improvements. Together we'll make positive changes and show quick results. 5S is where we'll start, and no matter what changes we make, we will always maintain the five S's."

It's one thing to force a behavior; it's another to remind people of a commitment.

- Pat Campbell, Director of Operations and Maintenance, Cape Canaveral Group

As a general rule of thumb, introducing good housekeeping in gemba *reduces the failure rate by 50 percent, and standardization further reduces the failure rate by 50 percent of the new figure. Yet many managers elect to introduce statistical process control and control charts in* gemba *without making efforts to clean house, eliminate* muda, *or standardize.*

- From *Gemba Kaizen* by Masaaki Imai

Gemba means "the workplace"
Muda means "waste"

5S Training

Friday, April 16 7:10 a.m.

"You are surrounded by clutter, disorganization, dirt and waste material all day long, but you've become so accustomed to it that you don't see it anymore," Roger said. "These conditions in your work area are creating non-value-added work for you and are hurting the quality, cost and delivery of our products."

Robin, Steve, Amy, Rob, and Donna, assemblers in the wall clock assembly area, were listening carefully.

"For instance," Roger continued, "do you have any tools, material, or equipment in your work area that you haven't used in a while and may never use?"

Amy raised her hand. "We have a big piece of

69

equipment in our work area that I've never seen anyone use. I don't even know what it's supposed to do."

"We've got a couple dozen boxes of unidentified plastic parts in our incoming drop area," said Steve. "Does anyone know what they are?"

"All I know is they don't go in our product and they've been there at least six months now," said Robin.

"I've got another one," said Amy. "There are a whole bunch of tools in our tool drawers that we don't use anymore, and half of them are broken."

"Good examples," said Roger. "How do those things create non-value-added work for you and hurt our quality, cost and delivery?"

"We're always having to walk around that big piece of equipment," said Rob. "I know it's only seven extra steps, but those steps add up when you do them thirty times a day, 280 days a year. I could be making clocks during that time."

"We also could use that space for something else," said Donna. "We expanded our building last year. We probably could have avoided spending that money by using our current space better. Maybe we could sell that piece of equipment or maybe another department needs it but doesn't know we have it."

"I'd like to get rid of those boxes of plastic parts," said Steve. "Every time I get more parts for our product,

I have to move them, and they look so much like our boxes that I have to waste my time checking the labels."

"Another department may have needed those parts," said Donna. "They probably already wasted time and money making new ones."

"How about the old and broken tools in your drawers?" asked Roger.

"You have to spend time digging to get the tool you want," said Steve.

"All the clutter we've been talking about makes it hard for me to concentrate," said Amy. "It's always in the back of my mind taking my attention away from my work."

"I know what you mean, Amy," said Roger. "Let's do something about it by implementing the first S, which stands for Sort. Sort means to remove all items from the workplace that are not needed for current production. This includes:

- Excess supplies and raw materials
- Excess tools
- Obsolete WIP
- Unneeded tooling and equipment
- Damaged items
- Excess furniture, shelves and carts

"We're going to do this now and I'm going to help you," said Roger. "If it's obvious that we don't need something, throw it in the dumpster out back*. If you think it might be needed somewhere else, put it on a cart and bring it to the holding area we've established near the back door. We'll give everyone a chance to look at that material before we get rid of it."

*   *   *

"Great job getting rid of all that clutter," said Roger. "There is a lot of empty space in your work area. How do you feel?"

"I feel good," said Amy. "My mind is clearer now too. That makes me feel better about getting back to work."

"Not so fast," Roger said. "I said earlier that you were surrounded by clutter, disorganization, dirt and waste material, and that they were causing non-value-added work for you and hurting the quality, cost and delivery of our products."

"We took care of the clutter," said Rob. "Now we

---

* If you are nervous about giving your people complete freedom to throw things away, put everything in the holding area until everyone gets a chance to look through it. You may also label each piece with a tag telling where and who it came from and on what date (this is called the "red tag" method). Alternatively, you can hold a "5S auction", where everyone in the area gathers around while you point out each piece. If no one claims it, dump it.

need to get organized, right?"

"You nailed it," said Roger. "While working, have you ever had to spend time searching for tools or materials?"

"Are you kidding?" said Steve. "Almost every day! I try to organize the tools after I use them, but by the time I need them again, they're gone. Who knows where the last person put them, especially on the other shifts."

"He's right," agreed Donna. "Everyone puts the tools in different places when they are done, if they return them at all. Some people hide the tools so they'll be able to find them again. It makes me angry, but I can't really blame them."

"We usually have to search for the materials we need to make our products," added Robin. "Our incoming parts are dropped off in a different place every day."

"Good examples of disorganization," said Roger. "How do those things create non-value-added work for you and hurt our quality, cost and delivery?"

"Mainly it's a waste of time to look for tools and parts," said Steve. "It's also bad for our morale because it's irritating."

"It hurts quality because if you can't find the best tool for the job, you have to make do with another tool that doesn't do the job as well," said Donna.

"When people hide tools, we think they've

disappeared," said Steve, "so we end up having to buy more. Sometimes people take tools home, but we don't even know about it. How can we know a tool is missing if there's no storage place for it?"

"As far as parts go," said Amy, "if we don't know where the material handlers put them, they can get lost, especially small parts. That delays shipments and costs more for new parts to be made and delivered."

"We also have parts we call 'floor stock', like screws and rivets and solder paste," added Robin. "They are delivered whenever we ask for them, yet we run out all the time. We never know when we're getting low, because they aren't stored at any specific location."

"The biggest cost of disorganization is all the conflict we have," said Donna. "Conflict among ourselves and with the material handlers."

"I'm glad you understand why disorganization hurts us," said Roger. "Let's get organized by implementing the second S, which stands for Straighten*. Straighten means to establish a place for everything near its point of use, and keep it there from now on. Here is what we will do:"

---

* The Japanese version of this "S" not only means neat and orderly, but a sense of an attractive if not beautiful arrangement of the area and the things in it. Thanks to George N. Wells CPIM for this contribution.

- Establish pegboards for tools at the point of use. For each tool, label the board and draw an outline of the tool. Color code each board and each of the tools that belong on that board.

- Keep all tool storage visible. Eliminate storage of tools in drawers or cabinets. Eliminate personal toolboxes. They violate this guideline.

- Establish and label areas for incoming and outgoing material either on shelves or on the floor.

"You know," said Robin, "if there is a place for everything in our work area, and everything is in its place, it will be pretty obvious if there is anything in our area that doesn't belong there. If we keep everything straightened, we may never have to sort again."

"Very good point," said Roger. "Here is one more point about straightening: Studies show that disorganization costs a person approximately 30 seconds of searching time for every five minutes of working time," said Roger. "That adds up to 43.6 minutes for every eight-hour shift, or a nine percent productivity loss. Now let's start straightening!"

*   *   *

"We've taken care of the clutter and disorganization in your work area," said Roger. "Are we done?"

"No!" they all yelled.

"What's left?" asked Roger.

"The dirt and waste material," said Donna.

"Right," said Roger. "How do dirt and waste material cause non-value-added work for you and hurt our quality, cost, and delivery?" asked Roger.

"Dirt can get in our products and cause quality problems," said Steve. "It can also get into our machines and tools and cause damage and friction."

"Liquid wastes on the floor can cause you to fall and get hurt," said Robin.

"Dirty equipment hides problems," said Donna. "For example, if a press is leaking grease, a seal might be broken, which will cause serious damage if not fixed quickly. However, if the machine is dirty, you might not notice the grease."

"It comes back to what I said before about clutter," said Amy. "If you have a habit of neglecting your work area, you are probably not making a quality product either."

"I agree with what you've said," said Roger. "Now let's work to implement the third S, which is Scrub. Scrub means that everything in the work area is free of dust, dirt, grease, and grime, and that all manufacturing by-products have been removed. Here are some guidelines:

- Floors are swept and dry
- Equipment is shiny and free of grease and dirt
- Tools are cleaned properly before being put away
- Products are free of dust and dirt
- Containers are in place to catch shavings before they hit the floor
- Windows are clean and clear
- Floors and equipment are repainted and polished as necessary

"Our goal for Scrub is to be as clean as an operating room," Roger said. "We won't get there today, but let's get started."

*   *   *

"Your work area looks great," said Roger. "You've made a fantastic start with the first three S's: Sort, Straighten, and Scrub."

"Yeah, but it won't stay this way," said Steve. "In a day it'll be a wreck again."

"How do we stop that from happening?" asked Roger.

"We need some kind of schedule for straightening and cleaning," said Robin.

"Exactly," said Roger. "The fourth S is Schedule. Schedule means to define daily responsibilities for what is to be done and who is to do it in order to maintain and continually improve the first three S's. You'll need to post this schedule in your work area and follow it every day so that your work area is left in perfect condition at the end of each shift. Ten minutes per person per day is all it takes not only to maintain, but also to improve on the first three S's. By investing these 10 minutes each day, not only will you save yourself the 43.6 minutes of searching, but also all the other quality, cost and delivery problems caused by the absence of the first three S's will be eliminated. Now, go ahead and put together a schedule and show me when you're done."

*   *   *

Name	Responsibility
Robin	Mop floor
Steve	Sweep floor
Donna	Ensure tools and materials are returned to storage locations, replace labels and colored tape as necessary
Amy	Remove or dispose of unnecessary items
Rob	Clean press and remove metal waste

"Good schedule," said Roger. "But aren't you each

going to get tired of doing the same thing every day?"

"We're going to rotate once every week," said Donna. "Each of our names is written on a card with Velcro on the back. We'll move the names at the end of each week. If two people want to trade jobs for a day or a week, they can."

"Excellent," said Roger. "Are you ready for the fifth and final S?"

"Lay it on us," said Rob.

"The fifth S is Score. Score means to frequently measure how well each work area is performing with respect to the first four S's. As the president of Accurate Clock Company, I personally will be doing the scoring once per month. 5S is that important. If we can't do 5S right, we won't be able to do anything else right either. Take a look at the scoring method we'll use (see next page). I will give each work area a score from 0 to 2 on each question. 2 means that the question can be answered 'yes' with no visible exceptions. 1 means that there are a small number of exceptions, and it is obvious that the team made an effort to conform. 0 means that there are numerous exceptions (see chart on next page)."

Questions				If not a 2, why not?
Operators can recite and explain the 5S's.	0	1	2	
Previous scores are posted, as are the reasons for imperfect scores.	0	1	2	
Walkways are clearly marked and at right angles.	0	1	2	
Storage areas are marked, labeled with name and standard quantities and area will accept no more than standard quantity.	0	1	2	
The floor is absolutely free of loose material.	0	1	2	
Cleaning materials are stored and readily available.	0	1	2	
Nothing is stored outside marked storage areas.	0	1	2	
A schedule is posted with clear responsibilities to maintain Sort, Straighten, and Scrub.	0	1	2	
There is nothing in the area that is not needed in the next 30 days.	0	1	2	
Nothing is stored in drawers.	0	1	2	
There are no personal tools or toolboxes.	0	1	2	
Nothing is placed on top of machines, cabinets, or horizontal surfaces.	0	1	2	

It is possible to retrieve any tool necessary to perform a job within five seconds.	0	1	2	
There is no rust, dust or buildup on machines or equipment.	0	1	2	
Equipment and machines are painted and paint is not scratched or scarred in any way.	0	1	2	
New ideas have been implemented to help us improve our 5S condition for next scoring.	0	1	2	

"I'll be back in 30 days to score you," said Roger.

CHAPTER SUMMARY

The 5 S's

- *Sort*: Get rid of anything not needed.
- *Straighten*: Establish a place for everything and keep everything in its place.
- *Scrub*: Keep everything very clean.
- *Schedule*: Establish a daily routine in each work cell to maintain and improve on the first three S's.
- *Score*: Regularly measure how well each cell is doing at maintaining and improving the first four S's.

The Sale

Friday, April 16   12:14 p.m.   Conference Room

"I just made a sale that will almost triple our wind-up alarm clock business," announced Evanson. 128,400 units a year, first delivery in two months. We normally sell them for $7.50 each, but we dropped the price to $6.75 for them. Our manufacturing cost is still $7.44. If we can drop our cost 20 percent to $5.95, this sale will result in an additional yearly profit of $102,720."

"In addition," said Maria, "the cost reductions alone will give us $126,317 more in profits on our existing orders."

"Will this make the bank leave us alone?" asked Lois.

"This sale is a great start," said Roger. "But we need

to take advantage of it by reducing wind-up alarm manufacturing costs by at least 20 percent during our first kaizen event. Frank Fowler doesn't think we can do it, so he's still letting Mortimer look for someone to buy us."

"What if we meet our cost reduction goal?" asked Kim.

"We'll get his attention for sure," said Roger. "Look, he wants all of the $11 million we owe him. He figures they'll only get $5 million if they sell now. He wants to believe us, but we've got to prove it to him."

"Then that's what we'll do," said Sam.

Kaizen Event Preparation

Monday, April 19   5:40 a.m.   Conference Room

"Let's start by reviewing our overall lean manufacturing implementation strategy," Roger said. "Here are the steps:

1. Teach everyone 5S and start implementing it everywhere in the plant.
2. Choose a product for a model line, to serve as an example or pattern for the rest of our products.
3. Implement work cells, continuous one-piece flow, and standard work in final assembly.
4. Build work cells in component fabrication and use kanban to connect them to final assembly. Reduce

85

setup time and batch sizes and implement preventive maintenance.

5. Apply everything we learned on our first model product line to all of our other products.

"We established our overall goals of 90 percent lead time reduction, 5,000 defects per million products produced, and 20 percent productivity improvements for the next two years," Roger said. "I'm leading the third of 10 scheduled 5S implementation sessions starting at 7:00 a.m., which is why we are meeting so early today. As you all have heard, Evanson made a big wind-up alarm clock sale for us, so that is where we'll focus."

"Great," said Sam. "So we'll use a kaizen event to implement work cells, continuous one-piece flow, and standard work in final assembly, right?"

"Right," said Roger.

"How do we start?" asked Kim.

"With preparation," said Roger. "Preparation is critical for a successful kaizen event. The first thing we need to do is define the scope of the kaizen event. Where does this event start and where does it end?"

"I would say it starts when all the component parts have been delivered to assembly, and ends when the product is packaged and ready for delivery to finished goods inventory," said Kim.

"Good," said Roger. "Next we must choose a team leader for the kaizen event. We need a person from outside the area we are improving so that he or she will bring fresh thinking to our improvement efforts. The leader also must know how to be a coach, how to get other people to think of improvements and implement them. We also need to choose a leader-in-training for this event for two reasons: first, we need more leaders to run more events, and second, the leader often needs help coordinating all the activities during an event. I will be the first team leader. Our first leader-in-training will be Tom."

"But I don't have the experience, and I don't understand this stuff very well," Tom said.

"You'll learn," said Roger. "Lean manufacturing is the strategy we're using to turn this company around, and you must understand and actively lead this effort with me*."

Tom shrugged and nodded weakly.

"Next," continued Roger, "the team leaders must

---

* Joseph C. Day, CEO of Freudenberg-NOK, spent 35 percent of his time working with kaizen teams during the first two years of lean manufacturing implementation. As a result, his company increased sales in four years from $200 million to $600 million with record profits. They have held 2,500 kaizen events in 15 manufacturing plants and involved 90 percent of their 3,500 associates. From *Becoming Lean*, by Jeffrey K. Liker.

study and understand the process in order to determine achievable breakthrough goals for the event in terms of quality, cost, and delivery. By applying my lean manufacturing experience to a process, I can usually picture how to improve lead time by up to 90 percent, productivity up to 50 percent, and quality up to 75 percent. Those are breakthrough goals, but they are achievable. You will develop the same kind of judgment after you've been through a number of kaizen events."

"How do you study an area?" asked Kim.

"I make products," said Roger. "I like to start at the beginning and have the operators teach me each operation. Then I have them watch me do it. Doing the job yourself is the only way to learn the details. Of course, I also look at the levels of inventory, the way the equipment and tables are arranged, the volume of product, the causes of defects, and the attitudes and skills of the people in the area. I don't have a 5S implementation session tomorrow, so Tom and I will spend the entire day learning about the area. We'll meet again as a management team tomorrow afternoon to discuss the goals."

*   *   *

Tuesday, April 20   4:45 p.m.   Conference Room

"Roger and I made lots of products today," said Tom, "and we studied the wind-up alarm clock assembly area. Based on Roger's lean manufacturing experience, we think the following goals are achievable:

- Reduce work-in-process inventory by 80 percent
- Reduce lead time (from fabricated parts to packaged product) by 90 percent
- Reduce defects by 50 percent
- Improve productivity by 30 percent

"We can accomplish these goals by implementing work cells, continuous one-piece flow, and standard work," Tom concluded.

"Good goals," said Roger. "Next we need to choose a week for the event. I suggest we start on Monday, April 26, which will give us time to finish our preparations."

"In the next few days," he continued, "we will need to meet with all the employees in the area to explain the kaizen event goals, what we will be doing during the kaizen week, and ask for their ideas for improvement."

"I like that idea," said Maria. "We'll collect their ideas and give them to the kaizen team to implement. That way, the people not on the kaizen team will have more buy-in to the changes in their area."

"Next, we will need to choose the kaizen team members," said Roger. "We've already got our team leader and team leader-in-training. A kaizen team should have five to 15 members, and approximately 70 percent of them should be non-managers and people from the shop floor."

"Why?" asked Maria.

"Because we want to use kaizen events as a way to give our shop floor folks a chance to make changes," said Roger. "When you get too many salaried people together they talk too much and the hourly employees get discouraged and stop participating. There are a lot more brains on the shop floor than in the front office, and we need to tap into them. The salaried people on the kaizen team will be coaches, not thinkers and implementers."

"That makes sense," said Maria.

"Let's start by assuming our first event will have 12 people," said Roger. "Counting Tom and me, we have room for at most two more salaried team members, and they should come from the management team to get you involved early. As salaried team members, I will expect you to serve as coaches to keep the sub-teams on track during the event, in addition to providing ideas and enthusiasm."

"I'd love to volunteer, but I'm too busy defending us against our creditors," said Maria.

"I understand," said Roger. "Sam, you should be one of the team members."

"Great," said Sam.

"Paul," said Roger, "as quality manager you should be on this team. The work cells will be taking responsibility for their own quality control."

"I don't think I'd be of much help," said Paul, his arms crossed in front of him. "I don't agree with this lean stuff. I haven't from the start, and you haven't listened to my objections."

Roger stared at him and paused. "You made a commitment to help us implement lean manufacturing when you signed the paper in my office. In order for you to stay with this company, you will need to suppress your disagreement from this moment forward and commit yourself to supporting this effort with every word you speak and action you take," Roger finally said. "Will you do that?"

"I have a right to my opinion," Paul snapped back.

"The decision to implement lean manufacturing at this plant was made when you hired me," said Roger. "I will tolerate no opposition to that decision from any employee of this company."

"I didn't vote to hire you," Paul responded. "And that's not a very nice way to talk."

"Nice isn't my goal."

"I thought you were supposed to be a big 'team' guy," said Paul. "People on teams are supposed to be nice to each other."

"People on teams are supposed to share a common goal," said Roger. "If you don't share our goal of implementing lean manufacturing, you can't be on the team."

Paul stood up. "These lean ideas will sink this company," he said. "And Roy feels the same way."

Roy turned to look at Paul, clearly angry, but silent.

"Will everyone but Paul please leave the room?" Roger asked.

They all rose and silently left the room. Roger closed the door behind them.

"Paul," said Roger, "you're fired."

"I figured that."

*   *   *

"Paul has chosen to leave the company," announced Roger. "Roy, as engineering manager, you would be an excellent choice for our last salaried spot on the kaizen team."

"As Paul said," Roy responded, "I'm not completely comfortable with lean manufacturing. It's the opposite of what I've believed for 30 years. But I'm willing to swallow my disagreement and do my best to help."

"That's all I ask," said Roger. "We need to fill the remaining eight slots with non-managers and people from the shop floor. It is often important to ensure maintenance support for a kaizen event, and because we will be moving equipment and electrical drops, we should have a maintenance technician as a team member. It is also helpful to involve a supplier or customer on the team, so we will invite one or the other for this event."

"I know a buyer I'd like to invite," said Evanson. "I told him what we're doing to improve, and he'd be interested in participating."

"Invite him," said Roger. "We have six spots left and they should all go to shop floor employees."

"Should we ask for volunteers?" asked Kim.

"No," said Roger. "Not for the first kaizen event. In every department in a plant, there are those hourly employees who influence their peers. Their opinions tend to become the opinions of most everyone else. We will identify the 'opinion-leaders' and ask them to be on the team. A properly managed kaizen event is a wonderful experience for most hourly employees, and we want people who will spread the word for us. We'll

93

choose four people from wind-up assembly and two people from the departments that fabricate parts for them."

"I know who most of the opinion-leaders are," said Kim.

"Me too," said Sam.

"I suggest you two put a list together," said Roger, "and run it by Roy too. Then we'll ask them to join the team. You must understand that team members are dedicated to the kaizen team for the entire week. Get someone to handle your regular job for the week, get rid of your pager, and tell everyone else to pretend you're on vacation."

"I assume that alarm clock production will be interrupted for part of the week," noted Kim. "I'll look at our inventory and ensure that we build enough product before the event to enable a week-long shut down."

"There will be some disruption to the schedule as we move equipment and experiment with different layouts," said Roger. "Plan on half the normal production from that area. In addition, I would like you to take on the role of acting manufacturing manager in addition to your duties in production control. Can you handle it?"

"I'll do it," said Kim.

"Roger," said Roy, "you believe we will get an

immediate productivity improvement from this first event, right?"

"Yes," said Roger, "at least 20 percent."

"There are 30 people assigned to wind-up assembly," continued Roy. "Let's say during the event we reduce that by six. You promised no more layoffs, so do you have a plan to reassign employees freed up due to productivity improvements?"

"Eventually," explained Roger, "they will be needed for production because we are going to grow, right Evanson?"

"Sure boss," said Evanson. "If you give me better quality, cost and delivery, I can sell as many clocks as you make."

"Once they've had time to cross train," continued Roger, "I will take the best six people out of wind-up assembly and assign them to our brand new 'kaizen department', reporting directly to me. They will work on improvement projects throughout the plant."

"Is there anything else we need to do to prepare for our first kaizen event?" asked Roy.

"Just a few more details," said Roger. "We need to have a carpenter available to build new tables and storage racks during the event."

"I know the perfect person," said Roy. "She's a quick worker."

"Fine," said Roger. "Make sure she's available. We also need to identify administrative support for the team. We'll use Tom's executive assistant, since he has nothing for her to do yet. She will order t-shirts and plaques as recognition for the team members, arrange for lunch to be brought in each day, reserve a conference room near the shop floor for training and discussion, and make sure we have supplies: flip charts, markers, pens, pencils, scratch paper, stopwatches, rulers, calculators, tape, scissors, tables and chairs for everyone, notebook computer, LCD projector, video camera with tripod, blank tapes, television and VCR, name tags, clipboards, masking tape, sticky notes, and tape measure."

"Sounds like you've done a few of these kaizen events," said Tom. "Anything else we need to do?"

"Two more things," said Roger. "We need to get approval for overtime to be used during the week as necessary. If we manage the event correctly, we won't need much of it. I hereby approve the overtime. Finally, dress casually for the event. You will be working on the shop floor and getting dirty. No suits or ties."

"How often will we do kaizen events?" asked Roy.

"Given the circumstances, we need to plan on doing another event as soon as this one is finished, and we'll start planning that one right now," said Roger. "Tom, you'll be the leader of the next event. It will be your responsibility to plan it. I will coach you."

"Where is this next event going to be?" asked Tom.

"Work with Kim and figure out which fabricated wind-up alarm clock component has the most quality problems and the longest lead time," said Roger. "That's the one on which we'll focus."

CHAPTER SUMMARY

Pre-Event Preparation Checklist

✓ Define the scope

✓ Choose a team leader

✓ Choose a team leader-in-training

✓ Study the process and determine achievable breakthrough goals

✓ Choose a week for the event

✓ Meet with all the employees in the area to explain the kaizen event goals, explain what the team will be doing during the kaizen week, and ask for their improvement ideas

✓ Choose eight to 15 members. About 70 percent of them should be non-managers and people from the shop floor

✓ Ensure maintenance support

✓ Involve a supplier and/or a customer

✓ Identify the "opinion-leaders" and ask them to be on the team

✓ Ensure that team members are dedicated to the kaizen team for the entire week

✓ Build product before the event to enable a week-long shut down

✓ Plan to reassign employees freed up due to

productivity improvements

✓ Have a carpenter available to build new tables and storage units

✓ Identify administrative support for the team

✓ Order t-shirts and plaques as recognition for the team members

✓ Arrange for lunch to be brought in each day

✓ Reserve a conference room near the shop floor for training and discussion

✓ Get approval for overtime

✓ Dress casually for the event

✓ Gather supplies: flip charts, markers, pens, pencils, scratch paper, stopwatches, rulers, calculators, tape, scissors, tables and chairs for everyone, notebook computer, LCD projector, video camera with tripod, blank tapes, television and VCR, name tags, clipboards, masking tape, sticky notes, and tape measure

It's heady and a little frightening to know that the boss has put part of his or her reputation into the subordinates' hands. It brings out the best in everyone. They aren't just getting the job done. They're making sure that the trust that's been placed in them is rewarded.

- Peopleware, by DeMarco & Lister

Kaizen Week Orientation

Monday, April 26   7:00 a.m.

"Here is today's agenda," Roger said, pointing to a flip chart. He looked around the room at the 11 members of the kaizen team. "First, I'll start by describing the types of ideas we can implement this week and how we will make decisions together. Second, I'll explain the lean ingredients we will be implementing over the next four days. Third, we'll brainstorm a list of improvement ideas, add them to the ideas we collected from the shop floor, and prioritize them. And fourth,

101

we'll define a small number of projects to complete this week, divide you into sub-teams, and get to work."

Kaizen Ideas and Decision Making

"For 80 years at Accurate Clock Company, we've depended on managers and engineers to do the thinking," Roger explained. "That won't work anymore because we have too many problems to solve and too many improvements to implement. This week we are asking you to be the problem solvers and improvement implementers. I am putting the future of this plant and my reputation as a manager in your hands. You did not ask for this, but I am asking it of you. I believe in you and trust you to do the right thing."

"So we can implement any ideas we want to?" asked Barney from wind-up assembly.

"As long as they meet three conditions," said Roger. "First, they must be low cost. This week is about using our brains, not our budget."

"How 'low cost'?" asked Barney.

"$500 or less," said Roger. "Second, the ideas must be implemented by the end of the day on Wednesday."

"Wednesday?" asked Amanda, also from wind-up assembly. "I thought we had until Friday to implement our ideas."

"I know from experience that we will need all day Thursday as well as Friday morning for last-minute adjustments and 5S," explained Roger. "The third condition is that the improvement idea must help us reach our goals. Do you remember what they are?"

"I do," said Ken from maintenance. "Reduce work-in-process inventory by 80 percent, reduce lead time (from fabricated parts to packaged product) by 90 percent, reduce defects by 50 percent, and improve productivity by 30 percent."

"Right," said Roger.

"What happens if we disagree about what ideas to implement?" asked Ron, a buyer from a major customer.

"We decide using consensus," said Roger. "Do you know what consensus means?"

"It means that everyone agrees," said Ron.

"No," said Roger. "Consensus means 75 percent agreement, but 100 percent commitment. If we wait until everyone agrees with every idea, we'll never do anything. I expect that everyone on the team will fully express their opinions, but if it is clear that the majority has reached agreement, go out and try it. I'd rather see you spend an hour experimenting with solutions than sitting around discussing what might or might not work. The ideas we implement this week don't have to be perfect — they just have to work."

103

Don't be too timid or squeamish about your actions. All life is an experiment. The more experiments you make the better.

- Ralph Waldo Emerson

"But what if my peers on the shop floor don't like what we implement?" asked Angela from wind-up assembly. "I don't want them blaming me."

"I have three points for you to consider," said Roger. "First, we will meet briefly with the other 25 wind-up assemblers on both Tuesday and Thursday afternoons to explain what we are implementing and ask for their feedback. That will surely save us from a few big blunders. Second, to offset the automatic resistance to change, and because you may not have data to immediately demonstrate the superiority of your ideas, I have explained to your peers that I am trusting your decisions, and I intend to stick with them unless the data later show that they are not working. When the other shop floor employees participate in future kaizen events, I will grant them the same trust. Third, we asked you to be a member of this team because you are a leader of your peers. Some people may criticize your decisions in the short term, but in the long term, you will be respected for the results you achieve."

"That makes sense," said Angela.

Lean Manufacturing Training

"The majority of our efforts this week will be focused on implementing three lean ingredients," said

Roger. "They are continuous one-piece flow, work cells, and standard work. We also will continue the 5S implementation we started when I worked with you two weeks ago."

"I know you explained the lean ingredients to us before when you trained the entire workforce," said Lenora from wind-up assembly, "but I could use a refresher."

Roger spent the next two hours explaining continuous one-piece flow, work cells, and standard work to the team.* When he was finished, he concluded, "I've given you a verbal explanation of the lean ingredients on which we will be focusing this week. I will give you more training out on the shop floor while we implement them."

Brainstorming

"We know that we will be implementing continuous one-piece flow, work cells, and standard work this

---

* We will not describe the lean training here because you have had a review in earlier chapters, and you will be reading examples of each lean ingredient later in this book. When doing lean training at the beginning of a kaizen event, be careful to train only on concepts you will be implementing. It makes no sense to train on all lean ingredients if they will not all be used, because if they are not used, they will be forgotten.

week," said Roger. "However, we will also have an opportunity to implement several other improvements. I am going to ask you for your ideas now. We will add them to the list of improvement ideas we collected over the past few weeks from the employees on the shop floor, then we will prioritize them to create our to-do list for this week."

Roger turned on the LCD projector attached to his notebook computer, and a spreadsheet appeared on the screen at the front of the room. The first column was labeled "improvement ideas".

"We are going to use a technique called 'brainstorming' today," said Roger, "and we are going to follow the brainstorming rules. Does anyone know what they are?"

"No criticizing or evaluating," said Roy.

"The more ideas the better," said Ken.

"Combine, modify, and build on previous ideas," said Sam.

"You are all correct," said Roger. "In addition, I want to hear some ideas that would be impossible to implement, or that are downright silly. Now let's hear your ideas!"

The team members yelled out ideas while Roger typed as fast as he could on the keyboard to enter them into the computer and display them on the screen. Many

If at first the idea is not absurd, then there is no hope for it.

\- Albert Einstein

of the ideas were about how to implement the lean manufacturing ingredients Roger had taught them, but some of the ideas focused on other areas that needed improvement. Once they ran out of ideas, Roger combined them with the ideas collected from the shop floor, numbered them, and printed out the entire list of ideas for each kaizen team member.

"We've got a big list of ideas," said Angela. "How do you expect to implement them all this week?"

"I don't," said Roger. "We'll certainly keep every idea on this list for later, but I want this team to choose only a few to implement this week. In order to make that choice, we will use a technique called 'multi-voting'. There are 127 ideas on the list. You can each vote for 20 ideas that you think we should implement this week. You have 15 minutes to make your choices, then we'll collect your votes and discuss the results to choose where we will focus our time this week."

Everyone bent over their papers and started marking their favorite ideas. After 15 minutes, Roger asked each person to report the ideas they chose. He tallied the votes on his computer in a column to the left of the improvement ideas. Then he sorted the list from most votes to least votes.

"Let's take a look at the top twenty vote-getters," said Roger, "and see if we can reach consensus on which ones we will have time to tackle this week."

The team discussed the ideas and decided to implement the 10 ideas in the chart below, in addition to establishing continuous one-piece flow, work cells, and standard work.

Votes	Improvement Ideas
12	Train operators to recognize defective parts by collecting bad parts for employees to pick out of a bin full of good parts.
11	Stop 100 percent quality inspections if team has not had any rejects for a week.
11	Grind assembly fixtures that are too tight, takes too long to remove part.
8	Find a different way to store faces so ink does not smear.
8	Improve lighting to make it easier to check for defects and assemble small parts.
8	Eliminate electrical cords and air hoses on the floor.
8	Establish one clock for shift change and break times.
7	Decrease the sound from the vibratory part sorter.
7	Install shock-absorbent mats to reduce stress of standing while working.
6	Develop a cross-training plan.
6	Use pre-mixed glue instead of mixing during assembly.

Establishing Sub-Teams to Work on Projects

"We've got our work cut out for us," said Roger. "I've divided our to-do list into three logical groups of improvement ideas, and I'm going to assign responsibility for implementing them to three sub-teams."

Roger projected the following chart on the screen from his computer:

Sub-Team	Members	Projects:
"Lean"	Tom - C Lenora Barney	• Establish continuous one-piece flow, work cells, and standard work. • Develop a cross-training plan.
"Quality"	Roy – C Amanda Kevin Anton	• Develop training to certify operators to recognize defective parts. • Stop 100 percent quality inspections if team has not had any rejects for a week. • Find a different way to store faces so ink does not smear. • Use pre-mixed glue instead of mixing during assembly.
"Morale"	Sam - C Angela Ron Ken	• Improve lighting to make it easier to check for defects and assemble small parts. • Install shock-absorbent mats to reduce stress of standing while working. • Grind assembly fixtures which are too tight, takes too long to remove part. • Eliminate electrical cords and air hoses on the floor. • Establish one clock for shift change and break times. • Decrease the sound from the vibratory part sorter.

"Any questions?" asked Roger.

"Yes," said Anton from stamping. "Why is there a 'C' next to Tom, Roy and Sam?"

"Tom, Roy and Sam are the coaches for each of the sub-teams," explained Roger. "They are responsible for keeping each sub-team focused and organized. The members of the sub-teams are responsible for doing the thinking and problem solving."

"So we split up and work independently?" asked Kevin from molding.

"Your primary responsibility is to implement the improvement ideas assigned to your sub-team," said Roger. "However, if you ever find yourself with nothing to do, come see me and I will find a way for you to help another sub-team."

"When do we start working on our assignments?" asked Ron.

"Now," said Roger, smiling. "But rather than running out there and working, I would like you to start by sitting together in your sub-teams and thinking about how to approach your list of improvement ideas. Make a plan so that you can complete implementation by when...?"

"Wednesday afternoon!" they all said.

"Exactly," Roger said. "Each morning we will meet here at 7:00 a.m. to spend a few minutes planning before we start working. In addition, we will meet for lunch at noon each day and get a verbal progress report from each sub-team. We will work through any conflicts between implementation plans, give each other ideas, and find out who needs help."

"Can we start now?" asked Ron.

"I have two more points to make to you," said Roger. "First, you are dedicated to this team for this week. If anyone wants you, tell them to pretend you are on vacation. Second, you will get frustrated some time during this week. It usually happens Tuesday or Wednesday. That is perfectly normal. Come to me and talk about it and we'll work through it. If we stick together we will accomplish great things this week, and set an example for the rest of the company."

"Now can we start?" asked Ron again.

"Go for it," said Roger

CHAPTER SUMMARY

The kaizen team may implement any idea that meets these criteria:
- Low cost
- Can be completed by the end of the day on Wednesday
- Will help the team reach its goals

Brainstorming rules:
- No criticizing or evaluating
- The more ideas, the better
- Combine, modify, and build on previous ideas
- Include ideas that might be impossible or downright silly

Kaizen decision making is by consensus:
- 75 percent agreement, 100 percent commitment

How to increase acceptance of changes by those not on the kaizen team:

- Meet briefly with non-participants on both Tuesday and Thursday afternoons to explain what the team is implementing and ask for their feedback.

- Management must announce their intention to stick with the kaizen team's decisions unless the data later show that they are not working.

Sub-teams

- Assign common sets of ideas to sub-teams, assign a coach to each sub-team.

- Have each sub-team create a plan to implement their ideas by Wednesday evening.

- Have each sub-team report their progress briefly to the entire kaizen event team each day at noon and resolve possible conflicts among improvement ideas.

- Expectation: If a sub-team member has finished his assignment, he should find a way to help another sub-team, or ask the team leader where he can contribute.

Factory Facts

Roger, Tom, Lenora and Barney gathered around a table to plan their approach for implementing continuous one-piece flow, work cells, and standard work in the wind-up alarm clock assembly area*.

"How do we begin?" asked Barney.

"We go out on the shop floor and determine our current quality, cost, and delivery," said Roger. "First, we'll measure first-pass yield, or the number of clocks that pass final inspection with no rework. That will be

---

* In order to provide examples of lean ingredients applied to the shop floor, we will only be following the activities of this sub-team during this kaizen week. The "quality" and "morale" sub-teams will be doing their work at the same time.

116

our quality measure. Then, we'll measure WIP (work-in-process inventory), productivity (clocks per assembly operator per day), part and operator travel distance, and space (the square footage used by assembly). Those are the cost measures. Finally, we'll measure delivery by finding out how long it takes to turn components into finished clocks."

"If we're going to improve all those measures," asked Tom, "why do we need to know what they are now?"

"I want to know how much better we got," said Lenora.

"That's right," agreed Roger. "We need to prove to ourselves and the rest of the plant that our work is worth the time and effort. After we are done, we will be able to estimate the financial savings and compare that to the cost of having this event. Then people will believe that we are making real change that's good for the business and good for them. That will make them more willing to help us turn this company around."

"Let's start at the beginning of the assembly process and walk through to the end," said Tom. "We'll collect our data and ask questions as we go."

"I've asked Kim to pull some of the data we need out of the computer system," said Roger. "She knows better than anyone how material flows. I'll call and ask her to meet us down there."

"Let's do it," said Barney.

*   *   *

"Assembly is divided into two parts," explained Kim. "The first is called the 'gear box assembly', located near receiving, where we assemble 63 gears, springs, bolts, nuts and other components into the interior assembly called a 'gear box'. That's where Barney works. The second area is called 'case assembly', located near the front office, where we add 23 exterior components including hands, face, glass, bells, housing, feet, a handle, and adjustment knobs. That's where Lenora works."

"How do the components get where they need to be?" asked Tom.

"Our MRP system tells us when to release parts to different areas, based on delivery dates and how long it takes us to do the assembly," said Kim. "Our batch size for wind-up alarm clocks is 500, so the system will print out a pick list of 500 of each component for gear box assembly. One of my material handlers will go to the warehouse and fill a cart with all the components we need to make 500 clocks. Normally, five to 10 components are either missing or short. Then the material handler puts all the components he collected on a shelf near the assembly area. Meanwhile, we start expediting from component fabrication or placing rush orders with our suppliers to get what we need."

"No wonder we're always late," said Tom.

"That's part of it," said Barney. "The real mess starts when we get bad components delivered to us. If one component in a batch is bad, many of the rest will be too. We waste time sorting through the entire box to separate the good from the bad, and then wait again for the parts we need. If the parts aren't in stock, we go back to expediting and rush ordering."

"In that case," said Kim, "they will assemble as much as they can, store the partially-built units on a shelf, and work on something else."

"And hope that the partially-built units don't get damaged before the parts arrive," added Barney.

"What happens to the gear boxes if they actually get built?" asked Tom.

"They go into inventory until MRP tells us to pull them out again, along with 500 each of the other 23 components needed for case assembly," said Kim. "In case assembly, we have all the same problems with short, missing, and bad parts."

"Who does the assembly?" asked Tom.

"In the gear box area," said Barney, "the supervisor gives an operator a blueprint of the interior of the clock and a list of parts. Then the operator has to go to the staging area to collect the components and bring them to an empty work table. That can take several days because the parts aren't organized and you have to search for them. You already know what happens when parts are missing or bad. It's the same in case assembly."

"Yes," agreed Lenora. "We spend more time trying to find parts than we do assembling them. Then during assembly, you have to find tools to use. The company keeps buying them but they keep disappearing. The only thing you can do is make your own collection and hide them when you leave, but you get in trouble for doing that. I went out and bought my own set. I couldn't find the same exact tools, but mine work okay."

"In gear box we have our own personal tool sets that the company buys for us," said Barney.

"When your supervisor hands you a blueprint, how often does it happen that you can actually assemble all the clocks for the batch?" Tom asked.

"Rarely," said Barney. "We end up standing around waiting for components or tools at least half of every day. Then when the components actually arrive, we work overtime to get the job out. All I want to do when I get here is be productive. I feel like all I do is waste time, and it's very frustrating."

"The other thing we do lots of is fixing mistakes," said Lenora. "If a clock fails final inspection, it comes back to be reworked. It takes twice as long to rework a clock as it does to make it the first time because you have to take it apart. Sometimes you can spend two weeks straight reworking one batch of clocks, if you have the parts you need to fix them. Otherwise, they sit and wait."

"Why do the clocks need rework?" asked Tom.

"Could be bad parts inside," said Lenora, "or improper assembly. Sometimes our department sends them back to gear box assembly for them to fix something."

"How often does rework happen?" asked Tom, who by this time was shaking his head and getting upset.

"I spend probably two days a week doing rework," said Lenora. "Every assembler does the job their way from the blueprints. People make mistakes."

"What's wrong with this company?" spouted Tom angrily. "This is insane."

"Tom," said Roger, "this is no different from most companies I've been in. Everyone is trying their best. The traditional manufacturing system within which they work creates the problems. That's why we're changing it."

"It needs to get a whole lot better very soon," said Tom.

"Let's collect our measures," said Roger. "Kim, how

much material has been released to wind-up assembly that has not been sent to finished goods inventory as clocks?"

"We've got enough material on the floor and in inventory to build 24,512 wind-up alarm clocks," she said.

"That's our work-in-process inventory," said Roger. "Can we use that number to figure out how long it takes to turn components into clocks?"

"Over the past three months, we've sold an average of 320 alarm clocks per day. If we stopped releasing alarm clock components to the floor today, we would use up all our WIP inventory in 24,512 divided by 320 equals 76.6 days. So that's our lead time."

"So why don't we stop releasing material to the floor for 76.6 days?" asked Tom, exasperated.

"In the current manufacturing system," said Kim, "we need all that material on the floor in order to have enough good parts in the right place to make 320 alarm clocks a day."

"When we implement lean manufacturing," said Roger, "we will do as you suggest. When we implement all of the lean ingredients for a product line, we will have only the parts we need when we need them. No more, no less."

"That sounds great," said Tom.

"That's what I've been trying to tell you," said Roger. "How about productivity?"

"We've got six people in wind-up case assembly," said Lenora.

"We've got nine in gear box," said Barney.

"And I figure we've got two full-time material handlers assigned to wind-up alarm clock assembly," said Kim. "Therefore, our current productivity is 320 clocks per day divided by 17 people equals 18.8 clocks per person per day."

"All we need now is first-pass yield, space, and part and operator travel distance," said Roger. He handed Lenora what looked like a short broom handle with a wheel on the bottom. "This will measure the distance you walk. I'd like you and Barney to go back to gear box, and wheel this device along the entire route that you and the clock parts take during assembly. That includes going to the warehouse and back and looking for parts and tools. Got it?"

They smiled and nodded.

"Also, while you are out there," continued Roger, "I'd like you to measure the length and width of the gear box and case assembly areas, along with any area used for wind-up alarm clock component storage. Then use those numbers to calculate the square footage of the assembly area. Find me if you need help. We'll meet in the training room when you are done. I'm going to check on the other sub-teams."

Lenora and Barney walked off toward gear box assembly.

"Tom," Roger said, "I'd like you and Kim to go to quality assurance and calculate our first-pass yield for the wind-up products. How many clocks are made right the first time, with no rework between gathering the components and the completed product?"

"We'll find out," Tom said.

*   *   *

"The part and operator travel distance was 2,350 feet," reported Lenora. "That's almost half a mile."

"The area dedicated to wind-up assembly and storage is 1,144 square feet," reported Barney. "Three-quarters of that is for WIP alone. Also, other departments have some of their material in our area, and we have some stored in theirs, so it equals out in the end."

"Very good," said Roger. "How about quality, Tom?"

"Not too good," Tom said. "We calculated that only 68 percent of the clocks get through assembly without needing some kind of rework."

"Well then, here is a chart of our starting point," said

Roger. "I can't wait to see where we end up:"

Measure	Pre-Kaizen
Quality (first-pass yield)	68%
Productivity (clocks per person per day)	18.8
Space (square feet)	1,144
Part and operator travel (feet)	2,350
Lead Time (days)	76.6

Continuous One-Piece Flow

Monday, April 26   3:55 p.m.

"Are you ready to implement continuous one-piece flow?" asked Roger.

"I'm ready for anything that will fix the disaster I just saw," said Tom. "I wasn't listening the first time you explained this stuff. Remind me, what is continuous one-piece flow, and why do we want it?"

"For this kaizen event, continuous one-piece flow means once we start assembling a clock," Roger explained, "it never goes back into inventory, and it never stops moving until it is a finished product, at which time it is sent to finished goods inventory. That means as soon as an assembly step is completed on a

126

clock, that clock moves immediately to the next operation which is performed immediately."

"And why do we want that?" asked Tom.

"Because with continuous one-piece flow there is no place for problems to hide," said Roger.

"What do you mean by that?" asked Lenora.

"I'll give you an example," said Roger. "How long would it take you to assemble an entire clock from components to finished product, including testing and inspection?"

Lenora and Barney discussed this question for a moment, and Lenora said, "Fifteen minutes."

"Okay," said Roger, "let's pretend we divided the job of assembling a clock into 15 one-minute assembly steps, and lined up 15 people in an assembly line to make clocks. Now let's pretend that the first person on the line made an assembly mistake or installed a bad part. If he did not notice the problem, when is the soonest the problem would be recognized?"

"By the next person in the line," said Barney.

"In that case, how many clocks need to be reworked?"

"One," said Barney.

"Right," said Roger. "When is the latest the mistake would be recognized?"

"At the last step, inspection and testing," Barney said.

127

"In that case, how many clocks need to be reworked?"

"Fifteen," said Barney.

"How many clocks do you rework now if someone made a mistake or installed a bad part during assembly?"

"Usually if one clock is bad, then all the clocks in a batch are bad," said Lenora, "so that means we'd have to rework 500 clocks."

"Actually," said Barney, "we'd probably have to rework every batch of clocks that person did, or every batch of clocks that contained that bad part. Kim said there are enough parts for 24,512 clocks on the shop floor now. If one of the first components in the clock assembly is bad, it might not be noticed until final inspection and test. In that case, we might have to rework every single clock on the floor."

Tom turned white and looked like he was going to pass out.

"Now you are starting to understand why your company has problems," said Roger. "The only reason you don't have more problems is you've been making this clock for so long. Just think how bad it would be if you were constantly making design modifications and introducing new components and assembly methods."

Tom was now making a gagging sound and holding onto the table to steady himself.

A line that never stops is either a tremendously good line or an absolutely terrible line.

- Taiichi Ohno as quoted in *Kanban: JIT at Toyota*

"With continuous one-piece flow, if a mistake is discovered, everyone stops working until the mistake is fixed permanently," said Roger. "The mistake can usually be fixed quickly because it was only made a few minutes before. In your current system, mistakes are fixed weeks or months after they happen, and everyone has forgotten how it happened. So the same mistakes keep happening again and again."

"That's us," said Lenora.

"With continuous one-piece flow," said Barney, "it seems like we'd spend much less time doing rework because we'd have to fix fewer clocks when a mistake gets made, and we'd never see that same mistake again. So productivity would have to go up."

"And lead time would have to go down," said Lenora, "because the amount of time it takes to turn components into a completed clock would only be fifteen minutes."

"And scrap would decrease also," said Barney, "because when we make a mistake that can't be reworked, we'd have fewer clocks to throw away."

"And our WIP would decrease also," said Lenora, "because we'd have components, 15 clocks being assembled, and that's it. No boxes of partially-built clocks in inventory or bad clocks sitting on shelves waiting for rework."

Tom was recovering and by now had some color in his face. "I'm convinced," he said. "So how do we implement continuous one-piece flow?"

"We set up a work cell," said Barney.

"Exactly," agreed Roger. "Do you remember the definition of a work cell?"

Barney smiled. "A work cell is a group of all the equipment and work stations necessary to make an entire product, or a large component of an entire product," he said. "They are placed as close together as possible so that product can be passed to the next operation as soon as the previous operation is completed."

"Outstanding," said Roger. "How do we figure out how many work stations we need in a work cell?"

"I have no idea," Barney said.

"Takt time," said Roger.

CHAPTER SUMMARY

- **Definition of continuous one-piece flow**: Once work on a product begins, it never goes back into storage, and it never stops moving until it is a finished product. This means that as soon as an operation is completed on a product, that product moves immediately to the next operation which in turn is performed immediately.

- **Why continuous one-piece flow?** No place for problems to hide. If a mistake is discovered everyone stops working until the problem is solved permanently. This is called line stop. The root cause of a problem can usually be fixed quickly because it was only made a few minutes before. In a traditional manufacturing system, problems are found weeks or months after they were created, and no one can remember how it happened. As a result, the same problems keep popping up again and again.

- **What are the benefits of continuous one-piece flow?** Rework and scrap decrease because there are less bad products to fix when a problem is discovered. Lead time improves because products don't wait in batches. WIP decreases because there is only one unit at each station.

- **How does one create continuous one-piece flow?**
  Set up work cells. A work cell is a group of all the equipment and work stations necessary to make an entire product, or a large component of an entire product. They are placed as close together as possible so that the product can be passed to the next operation as soon as the previous operation is completed.

Takt Time

"The first step to establish continuous one-piece flow is to determine how many clocks we need to produce each day in order to meet demand," Roger explained. "We will use that number to determine the rate at which our work cell will need to produce clocks. That rate is called the takt time. Takt is a German word meaning 'pace'. It is equal to the number of operating hours for the work cell divided by the daily demand."

"We work an eight-hour shift with a 30 minute lunch and two 10-minute breaks," said Lenora. "So our working time is equal to seven hours and 10 minutes, or 430 minutes."

"Let's make it 400 minutes," said Roger. "You'll need time for 5S and to implement improvements."

"Kim said the average demand for the last three months was 320 per day, so our takt time would be 400 divided by 320," Lenora continued as she punched buttons on a calculator. "That means our work cell will have to produce one clock every 1.25 minutes to meet demand."

"Excellent," said Roger. "Now I want you to think of a work cell as a short assembly line. If it takes 15 minutes to assemble an entire clock, how many assembly operators do we need to produce a clock every 1.25 minutes?"

"Well," Lenora said thoughtfully, "if I was the final person on the assembly line, I would need to finish a clock every 1.25 minutes. That means I would only have time to do 1.25 minutes worth of work on each clock – just the final inspection. The person before me would have to hand me a fully assembled clock every 1.25 minutes so I can inspect it. She only would have time to do 1.25 minutes of work, probably just putting the back and the knobs on. So if each person in the line does 1.25 minutes of work, the number of operators we need would be 15 minutes divided by 1.25 minutes, which equals 12 operators."

"Exactly right," said Roger. "Another consideration is that we will want to have each work cell operate as a team, like a small business within the business. Is 12

135

people a good sized team?"

"That's too big if we want them to make decisions without supervision," said Tom. "They'd need a referee to agree on anything."

"The perfect size would be four to seven people," said Barney. "That way there is enough diversity of ideas and opinions, but they'd still be able to make decisions independently."

"I agree," said Roger. "So what do we do about this 12-person work cell?"

"Divide it into two cells of six people each*," said Lenora.

"How would that work?" asked Barney.

"Each work cell would produce half of the daily demand," explained Lenora. "The takt time for each cell would be 400 divided by 160 equals 2.5 minutes. 15 minutes per clock divided by 2.5 minutes equals six

---

* Toyota Production System (TPS) adherents would want one work cell with 12 people for maximum flexibility and to allow for the removal of people as productivity improves. For example, if you improved the productivity of a 10-person cell by 10 percent, you could remove one whole person. If you improve the productivity of two five-person cells by 10 percent each, you can't remove a half person from each cell. I understand the TPS argument, but I think you will be better off keeping teams smaller to encourage ownership and accountability. You should take advantage of this structure by transforming your supervisors from babysitters into coaches who can develop your hourly employees' ability and willingness to think at work.

operators or work stations."

"Would they use the same equipment on two different shifts?" asked Barney.

"I don't know," said Lenora.

"That comes down to the cost of the equipment and how much space we have for the work cells," said Tom.

"Assembly equipment doesn't cost much," said Barney, "and we have the space we need for now. Most people would rather work days if possible."

"Then it's decided," said Roger. "Tomorrow you will design your work cells."

CHAPTER SUMMARY

- Takt is a German word meaning "pace". It is equal to the number of operating hours for the work cell divided by the daily demand.

- Each station in a work cell should be assigned the amount of work that can be accomplished within the takt time.

- If possible, each work cell should be staffed with four to seven people. This number encourages teamwork because it is small enough so that 1) each person has a large proportion of the accountability for results, and 2) the group can make decisions independently (without a supervisor acting as a referee). This is also large enough to provide for a diversity of ideas and opinions.

Work Cell Design

Tuesday, April 27   7:10 a.m.

"The next step in designing a work cell is to divide the 15 minutes of work necessary to make a clock into six 2.5-minute stations, one for each operator," said Roger. "Collect enough components to make 100 complete clocks, and divide the components into six stations on a row of tables on the shop floor. Then borrow some experienced operators from the shop floor, and make clocks."

"What if it turns out that one station takes less or more time than the rest?" asked Lenora.

"Then stop making clocks for a few minutes, and move the components from one station to another to

139

balance out the work," said Tom, "until each station takes the same amount of time."

"Right," said Roger. "Once you think you've got it right, move people around to different stations and do it again. That will give you more ideas."

"I'll be in my office," said Tom. "Let me know when you're done."

"No, Tom," said Roger, "you'll be on the shop floor making clocks."

"I will?" said Tom.

"Definitely," said Roger. "I'm going to work with the other sub-teams now. I'll meet you back here in a few hours."

*   *   *

10:22 a.m.

"Having fun?"

Tom looked up from the clock he was assembling and saw Maria looking down at him, hands on her hips. She was grinning.

"Actually I am," said Tom. "This is relaxing."

"Learning anything?" she asked.

"At first I didn't think I would," he said, "but it's a lot of fun thinking up ideas with Lenora and Barney and the others." He lowered his voice then said, "To be honest with you, I never had much respect for the people who work here, but now I'm starting to really like them."

She raised her eyebrows. "Better than your friends in New York?"

He frowned and looked thoughtful.

"Are you still planning on going back there?" she asked.

He looked up at her and hesitated. "I guess so," he finally said.

*   *   *

11:15 a.m.

"How did it go?" asked Roger.

Barney, Lenora, Tom and two other operators looked up from assembling clocks and smiled.

"Remember how we calculated that six people would be the right number for the work cell?" asked Barney.

"Yes," Roger said.

"Well, we think we can get it down to five."

"Five?"

"We're figuring out how to do it better," Barney explained. "None of us had ever assembled an entire clock, and the more we see of the assembly process, the more ideas we get."

"Wonderful work," said Roger. "You're ready for the next step of work cell design. I'd like you to rearrange the assembly tables until you've developed the ideal layout for the work cell. Normally work cells are arranged in a U-shape, with the operators on the inside of the U."

"Why a U-shape?" asked Tom.

"Because a U-shape minimizes the distance an operator needs to move to help another team member if there is a problem," explained Roger. "It also helps communications, because everyone is close to one another."

"Should the material flow clockwise or counterclockwise?" asked Tom.

"Is there a good reason for it to flow in one direction or another for this work cell?" asked Roger.

"Not that I can think of," said Tom.

"Then you can make the choice on your own," said Roger. "Direction can make a difference when you are using machines that load on one side and unload on the other, but you don't have that situation here."

"If we're going to make 160 clocks per day," asked Lenora, "where are we going to put the components? It's a mess having them all over these tables."

Roger pulled out his wallet. "Here is $500 in cash earmarked for this kaizen event," he said as he handed it to Lenora. "Take it and go to the hardware store and buy shelves and containers to hold your components. You'll set them up on your assembly tables within easy reach. Color code the containers for every station so they won't get mixed up, and label them with the part number and description. Buy two containers for each part. Each will hold a day's supply of components. That way the material handler will have time to fill up one container while you are using the parts from the second, and you'll never run out."

"That's called two-bin kanban, right?" asked Tom.

"That's right," smiled Roger. "You were listening after all."

"What about our tools?" asked Barney. "Where do they go?"

"Remember when I talked about 5S?" asked Roger. "What was the second 'S', and what does it mean?"

"Straighten," said Barney. "And it means 'establish a place for everything and keep everything in its place.'"

"Right," said Roger. "And where should the place be?"

"At the point of use," said Barney.

"You answered your own question," said Roger. "While you're at the hardware store, pick up some pegboard and pegs to make tool storage racks for each assembly station. Draw an outline in marker around each storage location on the pegboard, and label the location so that everyone knows where the tool belongs."

"We'll color code the pegboards the same color as the component containers for each station," said Lenora.

"And why not color-code the tools as well," added Barney, "so it's obvious where they belong?"

"Good idea," said Roger. "You are really catching on."

"We'll have to leave room for extra clocks in case one station gets ahead," said Tom.

"Tom," said Roger, "there won't be any extra clocks, because no station will ever get ahead. We're going to use a concept called 'kanban squares.' You'll use colored tape to make a square the size of a clock on your assembly tables between each station. When an assembler at one station finishes his assembly work, he'll put the clock in the square between him and the next station. He won't start on another clock until the assembler at the next station removes the clock he just put there."

"That way we never build up excess work-in-process," said Lenora, "and if someone is making a mistake, it limits the number of bad clocks that can be made."

"And if one station gets ahead, they should help the station that is behind," said Barney.

"That's right," said Roger. "One last tip: Make the work cell as small as possible. Do you know why?"

"It saves space that we can use to make other products," said Lenora.

"And it makes it easier to communicate with each other, and less walking if you need to help someone else," said Barney. "And in a work cell that uses machines like in component fabrication, it allows one person to operate multiple machines."

"And there is no room for extra work-in-process to pile up," added Tom.

"Now even you are catching on, Tom," said Roger.

CHAPTER SUMMARY

- **U-shaped cells**: Arrange stations of a work cell in a U-shape, with the operators on the inside of the U. The U-shape minimizes the distance a person needs to move to help another team member if there is a problem. It also helps communication because everyone is close to one another.

- **Parts**: Use shelves and shallow containers to hold small parts. Set them up on your assembly tables within easy reach. Color code the containers for every station so they won't get mixed up, and label them with the part number and description.

- **Tools**: Store at the point of use. Use pegboard and pegs to make tool storage racks for each assembly station. Draw an outline around each storage location on the pegboard and label the location so that everyone knows where the tool belongs. Color code the pegboards and the tools so it's obvious where each tool belongs.

- **Kanban squares**: Use colored tape to make a square the size of the product or component on tables or the floor between stations. When an operator at one

146

station finishes his assembly work, he'll put the clock in the square between him and the next station. He does not start on another clock until the assembler at the next station removes the clock he just put there. That way we never build up excess work-in-process. If someone is making a mistake, it limits the number of bad clocks that can be made. If one person gets ahead, he should help the station that is behind.

- **Small work cells**: Make the work cell as small as possible in order to:
  1. Save space that can be used to make other products
  2. Make it easier to communicate
  3. Reduce walking distance to help others
  4. Allow one person to operate multiple machines
  5. Eliminate space for WIP to accumulate

The purpose of standard work is not to stop progress; it is to preserve progress.

\- Michael D. Regan

Standard Work

Wednesday, April 28   8:19 a.m.

"The last three clocks came out of the cell at an average of two minutes and 28 seconds apart," reported Roger as he looked down at the stopwatch. "You are right at takt time with only five people."

"We're all working at a comfortable pace and none of us is ahead or behind," said Barney. "This is a fun way to work."

"I feel like I'm working with a real team," said Al, one of the assemblers helping with the experiment. "I was more productive today than I've ever been at work."

"That's because we've eliminated most of the non-value-added work," said Tom.

"Terrific job," said Roger. "Are you ready for the next step?"

"I thought we were done," said Lenora.

"You've designed a great work cell," said Roger, "but now we need to document it."

"Standard work," said Barney. "If we don't write down the job that each station has to perform, we'll lose all the good work we've done. People will go back to doing it their own way."

"That's right," said Roger. "You worked hard to develop a method for assembling wind-up alarm clocks which is resulting in the best quality, cost, and delivery we've ever had, and this is the way everyone must do this job."

"But what if we think of an idea to make it even better?" asked Lenora. "Because we probably will."

"I am going to hold each work cell accountable for their quality, cost, and delivery," said Roger. "If you think of an idea that sounds promising to your team, you can try it temporarily. If it improves your performance, you will change your standard work. The purpose of standard work isn't to stop progress, it is to preserve progress."

"Having the jobs documented also will help us train new people," said Al. "Until now everyone taught new people their own way of doing each job, thereby

There is always a best way of doing everything, even if it be to boil an egg.

\- Ralph Waldo Emerson

multiplying bad methods throughout the plant."

"How detailed does the standard work need to be?" said Barney. "Do we have to write down every time we move a finger?"

"No," said Roger, "although I have seen standard work documented almost to that level of detail. Some people say standard work should be documented so a stranger off the street could use the standard work sheets to do the job with no training."

"That's silly," said Tom. "That situation would never happen, and it would take forever to write it that way."

"That's my point," said Roger. "You should assume the person learning the job has a basic familiarity with the product and the components. Standard work for a station should fit on one side of one sheet of paper. Use a digital camera to take a picture of how the parts fit together at each station. Then put the picture in a word processing program and add arrows and text to explain how to assemble the parts. Be sure to include any tips or tricks that you've discovered to make the job easier."

"What do we do with the standard work when we're finished writing it?" asked Tom. "Put it in a notebook and store it?"

"Nope," said Roger. "You'll laminate it or put it in a plastic sleeve and then post it above the work station so it will be there when needed. If a work cell makes more than one model of a product, you can keep the standard work sheets above each station on rings, like a loose-leaf binder. You'd flip to the correct standard work when you change products."

"What computer should we use?" asked Lenora.

"Eventually, I hope to get a computer for the wind-up alarm clock assembly area," said Roger. "For now, use Tom's computer and ask his business assistant to help you with the typing. I'm off to work with the other sub-teams. I'll see you at lunch."

CHAPTER SUMMARY

- Near the end of the kaizen week, document the changes you've made, or people will go back to doing their work the old way. This documentation is called standard work.

- If an employee thinks of an idea after the kaizen week that sounds promising, a supervisor can give approval to try it temporarily. If it improves quality, cost or delivery, change the standard work to incorporate the new method.

- Standard work is required to train new people correctly.

- When documenting standard work, assume that the person learning the job has a basic familiarity with the product and the components. Standard work for a station should fit on one side of one sheet of paper. Laminate the page or put it in a plastic sleeve, and then post it above the work station so that it will be there when needed.

Cross Training and
Work Cell Management

Thursday, April 29   7:40 a.m.

"Your standard work sheets look great," said Roger. "They will be necessary to complete your final kaizen project, developing a cross-training plan. How will you start?"

"Why can't each person stick to one station and operate the work cell like an assembly line?" asked Barney. "What are the benefits of cross training anyway?"

"I can think of lots of benefits," blurted out Lenora. "First, if we are not cross-trained, we won't be able to

155

help each other. Second, doing one job all the time can lead to repetitive-motion injuries. Third, the more jobs we know, the better we'll be at thinking of improvement ideas and implementing them. And fourth, standard work can become dull, which will lead to poor morale. Rotating through the jobs keeps people interested."

"I see your point," said Barney, "but I was talking to some of the people in gear box assembly, and they don't like what we're doing at all."

"What did they say?" asked Tom.

"That we don't appreciate how long it takes to become a good gear box assembler," said Barney. "The normal career path starts you as a contractor in case assembly, then you become a regular employee, then after working here a long time you can move to gear box assembly. They already paid their dues in case assembly, and they don't want to go back."

"I've seen this before," said Roger.

"They also are saying that if you're a jack-of-all-trades, you're a master of none."

"Anything else?" asked Roger.

"They feel like they do the skilled work of making the clocks work, while case assembly does the unskilled work of making them look pretty."

"Barney, do you understand the business reasons for why we are going to cross train?"

"Yes," Barney said. "I agree with what Lenora said. We proved it all day yesterday."

"I'm going to talk to all the assemblers, both in gear box and case assembly," said Roger. "I am going to treat them all like adults and explain to them the business reasons we are doing this, and most of them will understand like you do. In either case, I will tell them that cross-training is a crucial part of how we are going to work here, and if they have any ideas for improving the work cell, I will want to hear them."

"What if they still don't like it?" asked Barney.

"Most of them will," said Roger. "In my experience, people end up liking the variety of working in different jobs. They like the team environment of the work cell and helping each other."

"One more thing we haven't talked about," said Tom. "Who is going to make sure all the components get replenished when the work cell runs low?"

"Good question, Tom," said Roger. "I will be assigning one additional person to each cell. That will allow each team to perform its own materials handling, give them flexibility to do cross-training, and give team members time away from the cell to work on improvement projects. I expect you to make a standard work sheet for materials handling and include that job in your cross-training plan. Go ahead and spend the next two hours putting together a cross-training plan, and then let's review it together."

*   *   *

"Tell me about your plan," said Roger.

"First, we need to choose, from among the assemblers who were not on the kaizen team, who is going to be assigned to each work cell," said Lenora. "Barney should be in one work cell, and I should be in the other. Then we need to spend tomorrow training people on one station each. We should let them stay at that station for the next two weeks to let things settle down."

"After that," said Tom, "We should have people learn the jobs before and after them so they can help when necessary. We are thinking of materials handling as the job between the last and the first person in the line."

"Eventually," added Barney, "every person will learn every job in the work cell. When we thought of that, we got another great idea."

"What's that?" asked Roger.

"We think the supervisor of this area ought to learn all the jobs too."

"You do, huh?"

"Sure," said Lenora. "The supervisor should be able

to demonstrate the process to prove the takt time can be achieved consistently and with high quality. Tom did it, and all our managers should do the same thing."

"I agree," said Roger. "In fact, as a consultant, I always start my assignments by working on the shop floor so that the hourly employees know that I understand what they are doing. I will expect that from all my managers."

"We had one more idea," said Tom. "We thought that instead of sitting at one station all day, each person could move from one station to the next and build an entire clock. We'd follow each other around the work cell and move every two-and-a-half minutes."

"Wonderful thinking," said Roger. "That method of working is called a 'rabbit chase.'"

"I would like having ownership of making an entire product," said Lenora, "but people would be upset because they'd have to stand all day instead of sitting."

"Actually," explained Roger, "standing is far better for your body than sitting, as long as you are moving around."

"That sounds great to me," said Barney. "My back hurts from sitting so much every day."

"Here is the cross-training chart we came up with," said Lenora. "Our goal is for each person in the work cell to learn one new job each month."

159

"We had another great idea while we were talking," announced Tom. "We're going to keep a bulletin board with the cross-training charts and other team information near the entrance of the work cell. Most important, we will post our current performance and goals for quality, productivity, and lead time. In addition, we'll have an area where we'll write down our improvement ideas, which ones we're currently working on, and which ones we've implemented. If we have ideas we can't do during the week, we'll save them for the next kaizen event."

"You are talking as if you are a member of this work cell team," said Roger. "We're still going to need you in your office to be CEO after this event."

"I know," said Tom, "but the CEO of a clock company should make clocks once in a while also."

They all laughed, but they knew he meant it.

"We're also going to put pictures and names of the work cell team members on the board," said Lenora.

"Good job," said Roger. "You've got tomorrow to finish up. See you in the morning."

CHAPTER SUMMARY

Benefits of cross training:

- If operators are cross-trained, they will be able to help each other.
- Doing one job all the time can lead to repetitive-motion injuries.
- The more jobs employees know, the better they will be at thinking of improvement ideas and implementing them.
- Standard work can become dull. Rotating through the jobs keeps people interested.

Cross-training progression:

1. Start by having people learn the jobs before and after them so they can help when necessary.
2. Eventually every person should learn every job in the work cell.
3. Finally, employees should cross train between work cells.

Supervisors must learn all the jobs in the work cell to show that the takt time can be achieved consistently and with high quality.

Staffing alternative: Instead of sitting at one station all day, assemblers can follow each other from one station to the next and build an entire product or component. This is called a 'rabbit chase.' Standing is far better for your body than sitting, as long as you are moving around.

Kaizen Tour and
Celebration Planning

Thursday, April 29   4:43 p.m.

"Great job this week," said Roger. "Here are the results you've achieved:"

Measure	Before	After
Quality (first-pass yield)	68%	98%
Productivity (clocks/person/day)	18.8	26.6
Space (square feet)	1,144	450
Part and operator travel distance (ft)	2,350	20
Lead Time (days)	76.6	2*

* There are up to two days of components in the kanban containers. This amount will be decreased after component fabrication kaizen events.

163

Everyone cheered.

"The 'Lean' sub-team implemented continuous one-piece flow, work cells, standard work, and a cross-training plan," said Roger, "and their work gave us immediate results. The 'Quality' sub-team not only developed the much-needed defect-recognition training program, but also found and successfully tested a new glue for assembly that will result in fewer defects and higher productivity. Finally, the 'Morale' team implemented six different improvements that will make work safer and less stressful for our people, which they will surely appreciate. Tom?"

"I put together a schedule for our kaizen tour tomorrow morning," Tom said. "I've spoken to all of you about it, and here is what we agreed to do:"

#	Presenter(s)	Improvements to Explain:
1	Guides: Anton Angela Ken Ron Roy Sam	• Begin the tour by explaining the kaizen event goals • Lead your group from station to station
2	Lenora	• Continuous one-piece flow, work cells, and standard work
3	Barney	• Cross-training plan, metrics, pictures, improvement idea log

4	Amanda	• Training program to certify operators to recognize defective parts • Show changes made to face storage containers to reduce smearing • Pre-mixed glue chosen to replace glue requiring on-line mixing
5	Kevin	• Improve lighting to make it easier to check for defects and assemble small parts • Install shock-absorbent mats to reduce stress of standing while working • Assembly fixtures ground to make part removal easier • Electrical cords and air hoses of correct length suspended from ceiling (removed from floor) • One clock established for shift change and break times • Noise from the vibratory part sorter decreased
6	Guides: Anton Angela Ken Ron Roy Sam	• Conclude the tour by showing the kaizen event results and answering questions as time allows

"We've got six stations, and we want to keep the tour to 30 minutes," said Tom, "so each station has five minutes."

"That's our takt time!" said Amanda.

"I'm so proud," said Roger.

"The tour is scheduled to start at 10:00 a.m.,"

continued Tom. "We've invited everyone in the plant, but only 60 or so will be able to join us. We'll send them through in groups of six so we should be done in under two hours. We're going to video tape your presentations at each station and put a VCR and TV outside the cell so that people can give themselves a tour later when they have time."

"We'll meet at 7:00 a.m. tomorrow as usual and spend a couple of hours cleaning the area and preparing our presentations," said Roger. "At 9:00 we'll practice the tour on ourselves."

"Didn't you say something about a celebration?" asked Roy.

"Yes I did," said Roger. "I've made 12:30 p.m. reservations for us at The Steak and Salmon Smokehouse. As much as you can eat, on me."

"Hooray!" they cheered. All but Tom. He looked grumpy again.

"Hold your horses," said Tom. "I admit the results we've achieved are incredible. But who says they're going to last? We're all excited, but how about a month from now when the novelty has worn off? Won't everyone slide back into their old habits?"

"Good question," said Roy.

"I've seen it happen before," said Sam.

They all looked at Roger and waited.

Follow-Up

"The kaizen team's job is to implement improvements, prove that they work, and document them as standard work," Roger explained. "It is the supervisor's job to uphold the new standards after the kaizen event is over."

"How would you know if the standards are not being upheld?" asked Roy.

"The first way is visual management," said Roger. "Visual management means designing an area in such a way that all you need to do is glance at the work cell, and it will be immediately obvious if something is wrong. The work cells are posting performance measures on their bulletin boards every day. We've rearranged the work flow into a logical order, set a specific takt time,

167

established a place for every tool, and standard work sheets are posted above each work station."

"You said visual management is the first way to ensure that the standards are being upheld," said Roy. "What is the second way?"

"We're going to meet regularly with the kaizen team and the supervision from the area after the event to review performance," Roger said. "I prefer 14 days and 30 days after the event, and then every 60 days for a year. During these meetings we will also follow up on any improvements that were started but not finished during the kaizen event. I like to keep those to a minimum. We have nothing left over from this week."

"What if the area is not performing up to standard?" asked Tom.

"Then we investigate to determine whether 1) the standard work is flawed, in which case we change it," Roger explained, "or 2) the supervisor is not doing his or her job, in which case we'll give training and counseling."

"We went from 17 people assembling clocks down to 12," said Lenora. "Where are the remaining five going to go?"

"I've spoken to Kim about that and we've got several answers," said Roger. "First, remember we've got enough parts for 24,512 wind-up alarm clocks on the floor right now. About half of those are partially-

168

assembled clocks and we'll need several people to finish assembling them the old way. We won't fabricate or order any more of those until we use what we've got. The other parts on the floor are unassembled components. We'll need people to restock them.

"Second, we can start sorting through all the parts in inventory and finished goods. Those areas are a mess, and they don't have the staff to do it themselves. Finally, once you and Barney have trained everyone in the work cells, we'll move you and several other top performers out of the cells to work on improvement projects elsewhere in the plant. You may even get a chance to support marketing's efforts to find new customers.

"Finally," Roger continued, "we started lean manufacturing in wind-up alarm clocks because Evanson in marketing told me that our sales will be increasing. Soon we'll start up a third work cell requiring six people."

"That's the growth we talked about," said Sam.

"We saved 694 square feet of floor space during this event," said Roy. "How do we make sure that space doesn't get filled up with junk?"

"It isn't all saved yet. The leftover work-in-process is still occupying half of that space, but it will be gone soon. Tomorrow morning we'll clear out everything we can," said Roger, "then we'll rope it off and put a sign inside saying, 'Reserved For Future Growth.' As we get rid of the WIP, the free space will grow."

"That makes sense to me," said Roy.

"Then go home and get some sleep tonight," said Roger. "We'll meet at 7:00 a.m. to clean up the area and practice the kaizen tour."

CHAPTER SUMMARY

- The kaizen team's job is to implement improvements, prove that they work, and document them as standard work. It is the supervisors' job to uphold the new standards after the kaizen event is over.

- **Two ways to ensure new standards are being upheld:**
  1. Visual management means designing an area in such a way that all you need to do is glance at the work cell and it will be immediately obvious if something is wrong. For example:
     - Post up-to-date performance measures
     - Rearrange the work flow into a logical order
     - Set a specific takt time
     - Establish a place for every tool
     - Document standard work

  2. Meet regularly with the kaizen team and the supervision from the area to review performance 14 and 30 days after the event, and then every 60 days for a year. This includes following up on any improvements that were started but not finished

during the kaizen event.

- **If the area is not performing up to the standards set during the kaizen event,** investigate to determine whether:
    1. The standard work is flawed, in which case it must be changed, or
    2. The supervisors are not doing their job, in which case provide training and counseling.

- **Ensure saved floor space doesn't get filled up** with junk by roping it off, and put a sign inside the area saying, 'Reserved For Future Growth.'

Celebration and Disappointment

Friday, April 30   12:33 p.m.

Roger stood and raised his glass. "A toast to all of you," he said. "You did the impossible and I'm proud of you. I appreciate the energy and brainpower you put into this week. You've shown our entire company what a group of committed people can do, and this is only the beginning. I salute you."

"Hooray!" they all yelled.

As they were all clinking glasses, Roger noticed Evanson standing in the doorway looking very serious. He wasn't supposed to be here. Roger excused himself and pulled Evanson with him into another room.

"What's wrong?" Roger asked.

"They cancelled," Evanson said.

"The wind-up alarm contract?"

"That's right."

"Any idea why?"

"They got a lower price from a company in Thailand. They were using us as leverage."

"How much lower?"

"Another five percent."

Roger stood and scratched his head for a moment. "I need to stay here for now," he said. "Can you set up a staff meeting for this afternoon at 2:00? I'll tell the managers here."

"Sure. What are we going to do?"

"I don't know yet."

*   *   *

2:10 p.m. Conference Room

They all stared at the conference table, trying to absorb what Evanson had just told them.

"So it's over," said Kim. "No more second chances?"

"Tom and I called Frank Fowler at Indiana First Trust to tell him what happened," said Roger. "He feels bad for us, but he said that given the circumstances, they are going to have to sell the company by the end of next week. Mortimer's got three interested conglomerates lined up and ready to come in here starting Monday to inspect the plant."

"There's nothing we can do?" asked Sam.

"I asked him that," said Tom. "He said that if we come up with something earth-shattering he'd listen, but no promises. He'd need to see that we're his best chance of getting his money back, and we can't do it by only cutting costs. We need to bring in new customers."

"So let's do that," said Lois.

"In one week?" asked Evanson.

"We don't have any choice," said Tom.

"Next week we're aiming our kaizen event at cases, the component with the longest lead time and worst quality," said Roger. "Once we've improved that area, we'll be able to sell our superior delivery and quality, not just low cost. Go out and sell assuming we can do that for them."

"I'll try," said Evanson.

*   *   *

6:55 p.m.

"You still here?"

Maria looked up to see Tom in her office doorway.

175

"I'm writing a letter to send to our vendors," she said as she handed him a sheet of paper. "Hopefully they'll understand our situation and keep sending us material even though we can't pay them right now."

He sat down across from her and skimmed the letter. "You don't even work here, but you sure act like it," he said. "Which is nice of you since you're one of those vendors that might not get paid."

"Your father was my first client," she said. "I feel like I owe it to him."

He nodded and looked at her.

"Tom," she said, "you've changed since you came back. You stopped wearing those suits, and you're more relaxed."

"I don't know why," he said, shaking his head. "We're fighting for our lives. But you're right, I'm having fun. I used to think Wall Street was where everything important happened. I'm realizing now that without thousands of companies like us, there wouldn't be any money for Wall Street to count and move. I feel good here. I feel like I'm home."

"I never thought I'd hear you say that," she said. "You've wanted to leave Indiana since the day I met you."

He smiled. "Are you planning on starving yourself tonight or are you going to eat dinner?"

"With you?"

"Why not?"

"Very persuasive argument," she said. "I suppose I do have to eat sometime. I guess I'll let you buy me dinner."

Fabrication Kaizen

Monday, May 3   7:00 a.m.

"This week we are concentrating on case fabrication," said Tom, this week's kaizen team leader. "We'll start by studying the shop floor to learn how the work currently flows, and to determine our current quality, cost, and delivery performance. Roger is observing the event and coaching me," Tom continued. "He will also challenge our thinking and instruct us as needed. Kim is our leader-in-training this week, and she will lead the shop floor study. Ready, Kim?"

"Let's do it," she said.

*   *   *

"The fabrication shop manufactures six components," explained Kim loudly over the noise of the stamping presses. "Bells, hands, handles, gear box retainer plates, backs, and cases. Assembly needs 710 cases per week for three different products: 320 for wind-up alarms, 270 for wall clocks, and 120 for stand-up 'grandfather' clocks. Cases are shaped like a short tube with rimmed edges. They require four operations: 1) stamping to cut the shape and punch the holes, 2) welding to turn the flat piece of metal into a tube, 3) forming to add the rim, and 4) polishing to clean and shine the surface*."

"Where is the stamping press?" asked Lenora from assembly, who did not know what a stamping press looked like.

"Right behind you," said Anton from stamping.

She looked at a large machine that was not moving. "How do they know when to make parts?" she asked.

"When the computer tells us," said Kim, "we deliver

---

* We are going to assume for this example that the processing time for each operation is the same for all case types. For instance, welding for a wall clock takes the same amount of time as welding for an alarm clock. If this is not the case in reality, you will have a different way of balancing the work between stations each time you change over to making a different component.

the material to press #2..."

"Which is down for repairs again," interrupted Anton, "so we'd run them on press #1. It takes us an hour and a half to set up the press, but then we can do the parts at 200 strokes per minute."

"Then where do the pieces go?" asked Lenora.

"We send them to the warehouse until welding needs them," said Anton.

"Pam, how long does it take to weld one case?" asked Kim.

"Well, I have to align each one on the fixture individually," Pam from welding said thoughtfully. "That takes about two minutes. Then the welding is quick, about 30 seconds. Taking it off is another minute. So, three and a half minutes each."

Lenora was furiously punching buttons on her calculator. Soon she said, "So to weld an average of 710 cases per day it takes 5.2 people."

"Actually, we have six people," said Pam. "Three on each shift. Some days are really busy and other days we run out of parts."

"Then they go back into inventory until we need to send them to forming," explained Kim. "How long does it take you to form each case, Tony?"

"About a minute and a half," he said. "We do them in batches of 1,500. Then the material handlers pick them up."

"They go back to inventory until I get them," said Nabil from polishing. "I can do a part in 40 seconds, and I do them all day every day. There isn't any setup because I place them on the fixture and go."

"Then the parts go back into inventory again until we send them to case assembly," said Kim. "We don't deliver to wind-up assembly any more. They come to us and pick up 320 each day for their next day."

"Let's collect our current performance measures," said Sam. "We're measuring quality by dollars of scrap per month. What was the average over the past three months?"

"$6,450 per month," reported Kim.

"Wow," breathed Tom. "What causes the scrap?"

"We get scrap from stamping, welding, forming, and polishing," said Kim. "We stamp in batches of 7,500. If something's wrong in stamping, all 7,500 parts can be bad. We had a situation six months ago where the parts were stamped wrong and then welded and polished. We didn't catch it until case assembly when the alarm hammer wouldn't fit through. We had to scrap them all."

"The later we catch the error," added Sam, "the more it costs because we've invested more labor into the part."

"Who did the setup?" Tom asked, his face red.

"It wasn't a setup issue," said Kim. "The die was

damaged and no one noticed."

"Another time three pallets of cases fell off a stack in the warehouse and most of them were damaged," said Anton. "I remember the rush to replace them for a late shipment."

"Many times it's scratches after polishing," said Nabil. "The cases sit around and get moved a lot."

"How about labor hours per case?" asked Sam.

"Our reports say 7.7 minutes per case," said Kim.

"Is that high?" asked Tom.

"It's 30 percent higher than what you'd get if you calculated it using the times that Tony, Anton, Pam, and Nabil just told us."

"So their estimates are off?" asked Tom.

"No, they're pretty close," said Kim. "But every time there is scrap, there is rework, and frankly the fabrication area is run pretty inefficiently. Production runs often are interrupted to do expedited orders. One day people are too busy, and the next they're sitting around waiting for parts. They've got to charge their time somewhere, and it gets charged to the job they're working on because they'll slow down so they don't run out of work."

"Feast or famine," said Lenora.

"Right," said Roger. "MRP supposedly plans everything in advance, but nothing ever goes as planned. As a result, no one knows whether things are right or wrong, ahead or behind."

"What's our work-in-process?" asked Sam. "How many clocks' worth has been released to fabrication that hasn't yet been released to assembly?"

"24,850 at $4.09 per unit equals $101,637," reported Kim.

"How did you get $4.09?" asked Lenora, always interested in the numbers.

"Labor hours per unit multiplied by hourly wage, plus the cost of raw metal, plus storage cost, plus the interest we would have earned on the money if we had it in the bank instead," said Kim.

"Knowing the WIP units," said Anton, "we can calculate lead time to turn raw material into cases in assembly as 24,850 cases divided by 710 clocks per day equals... anyone have a calculator?"

"It's 35 days," said Lenora, calculator in hand.

"Not only does long lead time mean lots of expensive WIP," said Roger, "but also while the parts are sitting in storage they can be damaged or even become obsolete when the product is redesigned. Then perfectly good parts become scrap."

"That's happened," said Kim.

"Who wants to measure space and travel distance?" asked Tom.

"I will," said Jezelle from stamping, Roland and

Garth from maintenance, and Arnie, the salesman from the metal supplier.

"You can all do it," Tom said. "Lenora has the measurement tools, and she'll show you how to do the calculations. Let's meet back in the training room in 45 minutes."

*   *   *

8:45 a.m.

"Roger and I have one goal for this team," Tom announced. "We would like you to implement kanban, one-piece continuous flow, setup reduction, and preventive maintenance to build a work cell dedicated to producing each day all the cases needed by assembly the following day*."

The group was silent as they tried to grasp the meaning of Tom's words.

"Here is a chart of our current performance," Tom

---

* An experienced kaizen event leader will know after brief study what can be accomplished in a week. In this case, Tom and Roger did back-of-the-envelope calculations before the event and decided that the team could cut setup time on the press in half and reduce the welding time which, when combined with other lean manufacturing ingredients, would achieve the results described.

continued, "and the results Roger and I think you can attain by the end of this week:"

Measure	Pre-Kaizen	Improvement %
Quality (scrap/month)	$6,450	50%
Productivity (minutes of labor per part)	7.72	30%
Work-in-process	$101,637	80%
Space (square feet)	670	50%
Part and operator travel (feet)	3,100	80%
Lead Time (days from release of metal to finished case)	35	90%

"In other words," said Jezelle, trying to understand, "as we are making the cases that assembly will need tomorrow, they will be using the cases we made yesterday."

"Precisely," said Tom.

"I need a refresher course on the lean manufacturing ingredients," said Garth, "I'm not sure how we're going to use them to get the results you're asking for."

"We're going to spend the next four hours talking about that," said Tom. "Then we'll divide you into three sub-teams to make it happen."

*   *   *

12:30 p.m.

"Here are the sub-teams on which we've agreed," said Tom, "and the projects you will be working on to reach our goals.

Sub-Team	Members	Projects(s)
Work Cell & Kanban	Sam - C Anton Tony Pam Nabil	Design work cell for fabrication of metal clock bodies. Move equipment as necessary. Test work cell. Establish kanban system to connect fabrication work cell to assembly work cell.
Setup Reduction	Kim – C Preston Roland Arnie Lenora	Reduce setup time on stamping press by at least 50 percent. Document new setup procedure.
Preventive Maintenance	Tom - C Garth Jezelle	Document and perform preventive maintenance procedure for all equipment used to produce cases.

"Now it's your turn," he continued. "Get together with your coach and take the next 45 minutes to plan your approach and have some lunch. Roger and I are here to help if you have any questions."

Work Cell Design in Fabrication

1:20 p.m.*

Sam, Anton, Tony, Pam and Nabil gathered around a table with their pizza in hand.

"Let me see if I've got the right idea about this work cell," said Tony. "We've got to put stamping, welding, forming, and polishing all in one small area, and then figure out a way to produce three different types of cases every day, for a total of 710."

"That's what I heard," said Pam.

---

* Chapters 24 through 27 occur during the same period of time.

"Impossible," said Tony.

"I agree," said Anton.

"Why?" asked Nabil.

"Where do I start?" asked Tony. "First let's talk about takt time, which equals the available working hours of the cell divided by the daily demand. We're supposed to use seven and a half hours on each of the two shifts, so that's 15 hours per day. If we're making three different cases each day, that's three times that we need to change over the stamping press. At one and a half hours each time, that equals four and a half hours of setup time alone."

"Hold on," said Nabil. "The setup reduction sub-team is supposed to cut that time by 50 percent."

"Yeah, right. I've set up that machine hundreds of times," said Anton. "I don't want to burst your bubble, Sam, but no way can they cut the time in half."

"Based on what Roger and I saw during our preparation," said Sam, "we think they can, and we will trust them to do it."

Anton shook his head. The group was silent for a moment.

"Fine," said Anton. "Setup time is the least of our problems. Let's say they actually reduce the setup time to 45 minutes, for a total of two hours and 15 minutes of changeovers per day. That leaves 12.75 hours of working time per day. Divide that by 710 cases and you get one minute and five seconds per case. Now if I understand takt time correctly, each station in the work cell has to take one minute and five seconds, right?"

"Right," said Sam.

"Our new stamping presses run at 200 hits per minute," said Anton. "They make over two parts every second. Meanwhile, welding takes three and a half minutes per part. There is no way this is going to work."

"Can the press go any slower?" asked Sam.

"Sure," said Anton sarcastically. "We can cycle it manually as slowly as you want, but you'd be wasting an awful lot of capacity. Besides, we only have two presses and if you dedicate one to cases, we won't have enough capacity on the other press to do the rest of the stamping."

"Our presses are less than two years old," said Sam. "What did we do with the old presses?"

"One of them is still in our warehouse," said Anton. "We can't sell it because it's too old and slow."

Sam looked at him silently.

"You're saying we should use it for this work cell?" said Anton. "But it's old technology. It isn't computer controlled, it only does 120 hits a minute, it's only 60 tons..."

"And?" asked Sam.

189

"I see your point," said Anton. "It'll do the job. We're lucky we couldn't sell it."

"In a work cell, speed and high-tech are often not important," explained Sam. "Many companies have old equipment they can use to build work cells. Another option is to sell a piece of expensive, new equipment in order to buy several pieces of slower, used equipment*. Let's take a moment now to tell the setup team our setup reduction goal of 45 minutes and which machine we want to use. We don't want them reducing setup time on the wrong press."

*   *   *

"Assuming the old press will work," said Pam, "we've still got the problem of welding taking three and a half minutes per case. I can't work any faster."

"Kaizen events are about working smarter, not faster," Sam reminded them. "We've got two choices for

---

* Once I consulted for a few days with a wire-harness manufacturer. They had recently invested in an expensive, automated super-machine that could (after a long setup time) cut, strip, and terminate wires very quickly. They "had" to keep it busy in order to pay for it, so almost every wire for every harness waited in line to go through that machine. Of course, if you are going to spend all that time setting it up, you might as well make a few extra, right? You get the picture. No flow, lots of waiting, lots of WIP. They would have been better off giving that machine to their competition.

welding. Either we find a smarter way to do it that doesn't take as long, or we divide welding into two or more smaller jobs."

Sam went to the notebook computer and typed for a few minutes. "Look at this chart," he said as he projected it onto the screen (see chart 1 on next page).

"Cycle time is how long each operation currently takes," he explained. "Takt time is how long each operation should take in order to produce the volume we need using one-piece continuous flow. We need to adjust the work content of each job until the cycle times equal the takt time."

"What if we made another welding fixture for each case type?" said Pam. "Then Anton could start getting the next case aligned while I weld the previous case."

"Sure, I could help," said Anton. "It takes roughly six seconds to manually cycle the stamping press. But aligning the case on the fixture takes two minutes and that would put me over the takt time by one minute and one second." (See chart 2 on next page).

"You're right, Anton," said Sam. "But Pam's got the general idea. What else could we do?"

"I don't think they put much thought into how they designed my welding fixture," said Pam. "We could speed up both alignment and detachment. As long as

Chart 1: Starting Point

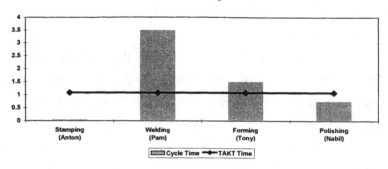

Chart 2: Anton helps Pam and new fixtures

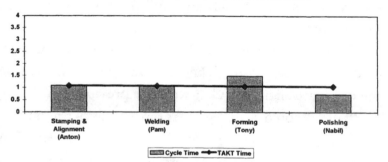

Chart 3: Nabil helps Tony

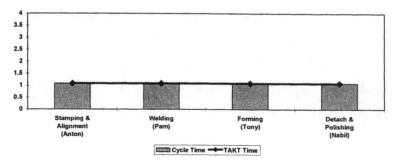

we're building another one, could we have input into how they design it?"

"Sure," said Sam. "I'll ask Roger to assign a tooling engineer to work with you and get it done this week."

Anton and Pam smiled. "That'll be fun," Pam said.

"For now, let's assume that will get you both down to takt time," said Sam. "Let me know tomorrow how you are doing so we can make another plan if necessary. Any other ideas?"

"I've got about 20 seconds after I'm done polishing and I could help Tony," volunteered Nabil.

"Another hand would help," said Tony. "I wouldn't have to walk around my machine to detach the case after I form the rim. I bet that would get me down to takt time." (See chart 3).

"Wonderful," said Sam. "We've got a plan. I'll get in touch with maintenance to help us move the equipment, then we can start testing our ideas. If we're wrong, we'll go back to the drawing board.*"

---

* This is one way to design this work cell. If the equipment was not needed full time to make cases, it would likely used to help make other parts as well. As the work cell members continue to reduce setup time, their goal will be to cut the batch size in half again and deliver twice per day to assembly. Another option (if the equipment is available) would be to dedicate a press just to making cases for wind-up alarm clocks. In this case, the fabrication equipment could become part of the assembly cell, producing cases one at a time as needed.

"Wait a minute," said Tony. "We're also supposed to be setting up the kanban system."

"Oops, you're right," said Sam. "Let's talk about that before we run out to the floor."

Kanban in Fabrication

"From what we learned this morning about kanban," said Pam, "it sounds like we need to build two containers for each type of case. Each container will hold one day's worth of parts. For example, the wind-up alarm clock containers would each hold 320 cases."

"So far so good," said Sam.

"During the day," Pam continued, "assembly will use the parts from the container we filled the day before. Meanwhile, we'll be filling the other container. At the end of their day, assembly will bring us the container they emptied so that we can fill it back up the next day. At the end of our second shift, we'll deliver a full container of cases to them so they'd be ready for the next day. We'll do the same thing for the other two product lines that use cases."

"Exactly," said Sam.

"We're also supposed to establish a specific place where assembly will leave the empty container," said Tony. "That way we won't lose track of it. Assembly is also supposed to establish a specific place for us to put our full containers."

"You learned well this morning," said Sam.

"Don't forget the kanban rule," said Tony. "If we don't have an empty container, we don't make the parts. We go find out what is happening in assembly and help them fix whatever is slowing them down."

"Okay, let's make sure we know our responsibilities," said Sam. "Anton and Pam, you are going to work on the new welding fixture, right?"

"Right!" they said.

"Nabil, will you work on the kanban containers?" said Sam.

"Sure," he said.

"And Tony," said Sam, "you and I will design a layout for the work cell and get help to move the equipment. We'll get input on the layout from the rest of the team first thing tomorrow morning, and aim to have the cell set up by noon. Any questions?"

They all shook their heads.

"We've got a couple more hours today," said Sam. "Let's do what we can for the rest of this afternoon and meet here first thing tomorrow morning."

The Prospect

1:22 p.m.

Tom's cell phone rang as he was gathering Garth and Janelle together to discuss their preventive maintenance plan. He excused himself to answer it.

"I just got off the phone with Ted Wittenberg, the V.P. of purchasing at West-Mart," said Evanson.

"West-Mart?" asked Tom.

"It seems they haven't finalized their contract for wind-up alarm clocks."

"And they're willing to talk to us?"

"He feels guilty about your father. I set up an appointment for us on Wednesday at 8:30 a.m."

"Have we got a chance?"

"I doubt it. We've got a terrible record with them, but we've got to try."

"I'm in," said Tom. "For something this important, the kaizen team can spare me for a few hours."

Setup Reduction

1:25 p.m.

"Our job is to reduce the setup time on the stamping press by at least 50 percent," said Kim, the coach of the setup reduction team.

Preston from stamping shrugged his shoulders and swallowed a bite of pizza. "It would be great if we could do that, but we can't," he said. "Anton and I have been changing that press for years, and we do it as fast as it can be done."

"According to Roger's training this morning, our setup situation isn't unusual," Kim said. "In fact, in his experience, most setup times eventually can be reduced to less than 10 minutes*. Who can remember the definition of setup time from our training this morning?"

"Setup time starts at the completion of the last good part from the previous setup and ends at the start of the first good part from the new setup," said Roland from maintenance.

"Right," said Kim. "Now does anyone remember the setup reduction process?"

"The first step is to video-tape the existing setup method and watch it to make a list of setup tasks and their durations," said Lenora from wind-up assembly.

"Yes," said Kim, "and we have to remember to set the camera so the elapsed time is recorded on the screen."

"Kim!" yelled a voice from the other side of the room. She looked and saw Sam waving at her. "Our team needs to talk to your team." Sam and his team came over and told them about their idea of using the older press, and the need to reduce setup time to 45 minutes.

---

* See *A Revolution in Manufacturing: The SMED System,* by Shigeo Shingo. SMED is an acronym for Single-Minute Exchange of Die, which means a setup time of less than 10 minutes. I know "single-minute" seems to suggest one minute or less, but it's a Japanese translation issue. Nevertheless, this is the best book out there on setup reduction.

*　　*　　*

"Now I've heard everything," said Preston. "If you want to go ahead and use that old press, I'll do what I can do to help, but you're all crazy."

"Great," said Kim. "We've got two hours left today. Let's get our videotape made so we can analyze it tomorrow morning."

"Before we do that," said Preston, "Roland and I should move that old press to where we're setting up the new work cell. We'll need Kim's clout to help get it done fast. Then I'll stay over during second shift so we can make the video."

"That sounds good," agreed Kim. "Lenora and Arnie, can you help Sam's sub-team this afternoon with the layout for the work cell?"

"Sure," they said.

"Let's do it," said Kim.

*　　*　　*

Tuesday, May 4　7:33 a.m.

"Two hours and five minutes," said Preston, "and most of the time I was either busting my hump or waiting for something outside of my control. There is no way we're going to get that down to 45 minutes."

"It doesn't have anything to do with how hard you worked," said Lenora. "Like Kim said before, most of our improvement is going to come from better organization and preparation, and some creativity. We have to work smarter, not harder."

Kim turned off the VCR. They had finished watching the video Arnie had taken of Preston changing the stamping press earlier that morning.

"We made a list of all the setup tasks and their durations while we watched the video," said Kim. "What is our next step?"

"The second step of setup reduction is to distinguish between internal and external setup tasks*," said Roland.

"And what is the difference between internal and external tasks?" asked Kim.

"A setup task is internal if it can only be performed when the machine is shut down," said Arnie, the salesman from the metal supplier. "A setup task is external if it can be performed when the machine is

---

* In reality, you will distinguish between internal and external activities while watching the video, but I'm showing them one after another to teach the concept. Take your time watching the video and allow people to take notes and ask questions. Type your list into a spreadsheet, recording a description and time for each activity.

running."

"Does anyone remember from our training why that difference is important?" Kim asked.

"With better preparation and organization, external tasks can be eliminated or done before shutting down the machine," said Lenora. "That alone can reduce the average setup time by 30-50 percent."

"Let's look at the list of setup tasks we made and decide which tasks could have been done before beginning the setup," said Kim. "We will make a checklist of all these tasks so that the press operator can ensure they get done."

"We need to make sure the die for the next part is located near the press and ready to go," said Arnie. "Preston had to spend 42 minutes waiting for the forklift and for repairs that should have already been completed."

"Excellent," said Kim. "What else?"

"He also spent lots of time looking for tools and parts," said Lenora. "It's hard to tell how much because it happened so many times. We need to make a shadow board and label a space for every tool needed to do the changeover*. That way we can see at a glance if anything is missing long before the next setup."

---

* Setup and changeover are synonymous.

"Also," said Roland, "we could get the material for the next part before we start the setup."

"Hold on," interrupted Preston. "When am I supposed to have time to do all this?"

"Remember," said Lenora, "you have three other people in the work cell who can help you. They will be finished with their work a few minutes after you shut down the press. Not only can they help you do the current changeover, they can also help you get organized for the next one."

"I guess that makes sense," Preston said.

"Good," said Kim. "We've saved a substantial amount of time by identifying tasks that can be done before starting the setup. What is the next step of setup reduction?"

"The third step is to convert internal tasks to external tasks," said Roland.

"I know something we could do there," said Preston. Everyone looked at him in surprise. "I spend at least 10 minutes adjusting the shut height of the press every time I change a die because every die is a different size. What if we welded some shims or metal blocks to the bottom of the dies so they would all be the same height?"

"That would require several different sizes of clamps to secure the bottom of the die to the press," said Roland.

203

"So we weld blocks on the inside of the bigger dies," said Preston. "That way we could use the same clamps for all dies that run on our press."

"Wow," said Roland, "That would work. That would eliminate most of the adjustments."

"Let's give it a try," said Kim. "What is the final step of setup reduction?"

"The fourth step is to eliminate or streamline all setup tasks," said Lenora.

"I've got another idea," said Arnie. "Remember I suggested keeping the next die near the press? Let's make a cart for each die with wheels on the bottom of the legs so we can move it easily. We'll make them the same height as the press bed and mount rollers on the top of them. That way we can roll the dies in and out of the press. We won't need the crane or the chains anymore."

"The other thing I noticed is that I have to walk around the press at least five times during the changeover," said Preston. "If I could train someone else in the work cell to help me, that would save some time."

"You work second shift," said Kim, "and so does Pam, who is working with the work cell and kanban sub-team. Maybe we can borrow her this week so we can try your idea."

"She could learn what I need her to do," Preston

agreed.

"The fifth step is to document and post the new setup procedure as standard work," said Kim. "We can worry about that on Thursday. Anyone remember the last step of setup reduction?"

"The sixth step is to measure and record every setup time," said Lenora, "because what gets measured gets done."

"Very good," said Kim. "We have enough ideas now to keep us busy. We are supposed to be done with implementation by tomorrow night. Arnie and Roland, let's have you two make the die change carts. Preston, you can handle welding the shims and blocks to the dies to eliminate setting the shut height. We'll get Pam to work with you first thing tomorrow morning. Lenora, you can design the shadow board for the setup tools and make pre-setup checklists. Make sure to check with Preston on the details. Does that sound good to everyone?"

"Sure!" said everyone but Preston.

"Preston, do you have another idea?" asked Kim.

"Our ideas will work," said Preston, "except for one reason. As I understand it, our work cell will only be able to use that one press, right? What if it breaks down? In the past we'd use a different press."

"That's not an option in a lean manufacturing plant,"

said Kim.

"Then our press better never break down, or we're dead," said Preston.

"That's why we have a sub-team working on preventive maintenance," said Kim.

CHAPTER SUMMARY

Definition of setup time

Setup time starts at the completion of the last good part from the previous setup and ends at the start of the first good part from the new setup.

Setup reduction steps:

1. Video-tape the existing setup method and watch it to make a list of setup tasks and their durations.
2. Distinguish between internal and external setup tasks. *A setup task is internal only if it can be performed when the machine is shut down. A setup task is external if it can be performed when the machine is running.*
3. Convert internal tasks to external tasks.
4. Eliminate or streamline all setup tasks.
5. Document and post the new setup procedure as standard work.
6. Measure and record every setup time.

Preventive Maintenance

"Our project is to document preventive maintenance procedures for all equipment used to produce cases," said Tom. "Do you remember from our training this morning why preventive maintenance is so important in a lean manufacturing plant?"

"After we set up work cells, we won't be able to share equipment anymore," said Jezelle. "Each work cell will have only one of each type of machine, and if that one stops working, the whole cell stops working. If the cell stops working, we aren't making parts and we aren't delivering to our customers on time."

"The other reason is that machines last longer if you do preventive maintenance," said Garth. "Breakdowns are tough on machines. For example, if dirt gets between two moving parts, it creates friction which makes the motor work harder. Friction also wears down the parts and makes them not fit together as well. Eventually the machine seizes up and parts break. Then it's a real mess to fix. We have to tear the machine apart and wait for new parts."

"And when a machine isn't working quite right, it isn't making quality parts either," said Jezelle.

"You've got me convinced," said Tom. "Do you remember the steps we need to follow to set up a PM program for our cell?*"

"The first step is to identify each piece of equipment in the cell," said Garth. "That's easy to do."

"The second step is to document the PM procedures for each piece of equipment," said Jezelle. "We can get that from the operating manuals or by calling the manufacturer of the equipment."

"The third step is to develop PM schedules for each piece of equipment," said Garth. "The schedules could be based on time (every two weeks), usage (every 20,000 strokes), testing (ability to hit a certain tolerance), or visual inspection (the presence of oil leaks)."

"The fourth step is to turn the PM procedures into easy-to-use checklists and post them near the machines,"

---

* Adapted from *Making Manufacturing Cells Work*, by Lee R. Nyman.

said Jezelle.

"How do we make sure all of this PM gets done?" asked Tom.

"The fifth step is to develop measures and charts to show scheduled versus completed PM activities," said Jezelle.

"Who is going to do all this preventive maintenance?" asked Tom.

"At first I will," said Garth, "but soon we want the cell owners to do it. The sixth step of developing a preventive maintenance program is for me to involve cell team members in all preventive maintenance and repair activities. First I'll show them how to do it. Then I'll watch and give them corrective feedback until they can do it without me. Eventually I'll only be there if they need me for special circumstances."

"We should be able to build many PM activities into our daily 5S schedule," said Jezelle.

"Good thinking," said Tom. "We need to get steps 1 through 5 done this week. Think we can do it?"

"Sure!" they said.

"Let's get started then," said Tom.

CHAPTER SUMMARY

Why preventive maintenance is important:

- Work cells don't share equipment.
- Machines last longer.
- When a machine isn't working quite right, it isn't making quality parts.

Steps to set up a preventive maintenance program:

1. Identify each piece of equipment in the cell.
2. Document the PM procedures for each piece of equipment.
3. Develop PM schedules for each piece of equipment.
4. Turn the PM procedures into easy-to-use checklists and post them near the machines.
5. Develop measures and charts to show scheduled versus completed PM activities.
6. Involve cell team members in all preventive maintenance and repair activities.

The Meeting

Wednesday, *May 5 8:44 a.m. West-Mart Headquarters*

"Sorry about your father," said Ted Wittenberg.

"Thank you," said Tom. "He'd been under a lot of stress and hadn't been taking very good care of himself."

"Ted," said Evanson, "you wouldn't have agreed to talk with us today if you were entirely satisfied with your current wind-up alarm clock supplier."

"For the past 20 years," explained Ted, "cost was our only criterion for choosing vendors. Recently we've learned that unreliable delivery and poor quality from our suppliers costs us millions of dollars each year in paperwork and unsatisfied customers. We'd actually be willing to pay a vendor more if they could reduce those

costs for us."

"Your current wind-up alarm clock vendor can't do that for you?" asked Evanson.

"No," said Ted, "but neither could Accurate Clock Company. Your father didn't understand our needs, Tom. He thought it was all about cost."

"What if we could give you the quality and delivery you're looking for?" asked Evanson.

"Frankly" said Ted, "I don't have any confidence you can do it."

"We can prove it," said Tom. "Visit our plant and see for yourself."

"I have a really busy schedule," said Ted, "but I feel I owe it to you."

"How's Friday?" asked Evanson.

Ted turned and leafed through his calendar. "I can be there at 6:00 p.m. Is that too late?"

"I hope not," said Tom.

"Pardon me?" said Ted.

"That's perfect," said Evanson.

Fabrication Kaizen Wrap-Up

Thursday, May 5   3:11 p.m.

"You built a work cell, you slashed setup time and welding time, you established one-piece flow, and you developed a preventive maintenance program," said Roger. "Then you spent yesterday making cases to prove that your ideas would work. Here are the results we've calculated (see chart on next page):

They all clapped loudly. Tom stood and waited until he got their attention.

"You all know we have a kaizen tour scheduled for 10:00 a.m. tomorrow morning," he said, "and then a celebration lunch. What you don't know is that we're giving another tour for West-Mart tomorrow at 6:00

p.m., and your work cell is going to be on center stage. Make sure the area is spotless, and that we've worked out any remaining bugs. Our future depends on it."

Measure	Before	After	Improvement
Quality (scrap/month)	$6,450	*	*
Productivity (minutes of labor per part)	7.7	4.5	41%
Work-in-process	$101,637	$3,479	97%
Space (sq ft)	670	350	48%
Part and operator travel (ft)	3,100	16	99%
Lead Time (days)	35	1	97%

* The team will need to collect data over the next three months to establish the new performance level.

The Decision

Friday, May 6   5:44 p.m.   Main Conference Room

"Gentlemen, you've seen everything you wanted to see," said Mortimer. He was smiling from ear to ear. He looked around the table at the executives from Acme Products Company. "We've agreed on a price of $4.6 million. You'll take the brand names, product designs, and customer list. We'll keep the real estate and sell it at auction."

Maria quietly moved to the door and slipped out of the room. Sam was waiting for her. "You sure the fire department won't show up?" she asked.

"Nope, we cut the line. We'll reconnect it later."

"You better do it now."

Sam nodded, then jogged down the hallway and disappeared around the corner. Maria slipped back into the room and took her seat.

"If there are no more questions," said Mortimer, "let's get this contract signed." He pulled five bound sets of papers out of his briefcase and lined them up neatly on the conference table, then produced a fountain pen and handed it to a heavy, balding man across the table.

When the fire alarms erupted everyone in the room jumped. Seconds later the sprinkler system came to life and started drowning the room with waves of water. For a few moments there was nothing but yelling. Suddenly a door opened on the far side of the room and a panicky voice screamed, "Over here! Hurry!"

The men in the room ran over each other to make it to the door, pushing one another into the walls and onto the conference table. Mortimer led the way down a short hallway and through another door that opened to the parking lot. A white van was parked directly in front of the exit, side door opened. A young woman was at the wheel. "Get in," she said. "I was told to take you to a safe place where you can dry off and change."

The men piled into the van. Kim Nguyen drove them out of the parking lot and down the road as Maria watched from inside the exit door.

217

*   *   *

"You can turn off the alarms and the water now, Sam," said Tom into his phone. He hung up.

Lois appeared in the doorway. "How long have we got?" she asked.

"Kim will keep driving until they force her to stop," said Tom. "We've got at least an hour. How is the rest of the plant?"

"Everyone's perfectly dry and at their work stations," she said.

*   *   *

6:10 p.m.   Front Lobby

"I've only got an hour," said Ted Wittenberg. "Show me what you've got."

"Follow Roger," said Evanson. Roger, Ted, Evanson, and Tom headed down the hallway and onto the factory floor.

"This place looks much different from the last time I visited five years ago," said Ted.

"We've made some changes," said Tom.

"Mostly it's cleaner," said Ted. "And neater. There used to be junk everywhere."

218

A moment later they were at the wind-up alarm clock assembly cell.

"Wow," said Ted. "They are working so efficiently, like..."

"Clockwork?" asked Evanson.

"Exactly," said Ted.

"We only have a few clocks in process at any one time," said Roger, "so if we encounter a quality problem we can stop and fix it immediately."

"Not only that," said Tom, "but each person in the work cell checks the work of the person before her."

"That will reduce your defects considerably," said Ted.

"Right," said Roger. "Follow me back to our fabrication shop."

The group followed him for another minute.

"This is our case manufacturing work cell," said Roger. "Cases were our longest lead-time component, not to mention the part that contributed the most defects. As you can see, we're now making them using one-piece flow, so we'll catch problems quickly, like in assembly."

"Most of our past delivery problems were the result of quality problems," explained Evanson. "We would discover too late that every part in a batch was bad, and we wouldn't be able to ship your clocks for the lack of one part. We won't have that problem any more with cases, and soon we'll manufacture all our components the same way."

They were interrupted by ringing coming from Tom's cell phone. He stepped aside and answered it. "A Mr. Frank Fowler is here to see you," said the receptionist's voice. "He seems to think we've had a fire."

"I'll meet him in my office."

*   *   *

6:40 p.m.   Tom's Office

"What is going on here?" asked Frank. His face was red and he was breathing hard, as if he'd been running. "I just got a call from Mortimer. He said your building was on fire."

"Where is he now?" asked Tom.

"For some reason he was 20 miles down the road toward Indianapolis when he called me," said Frank. "Said he'd meet me here any minute now."

Tom nodded and looked out his window at the parking lot.

"What's this about a fire?" Frank asked, looking calmer now. "Everything seems normal."

"We had a false alarm and one of our sprinklers leaked a little."

"Mortimer was talking as if the place had blown up." He was chuckling now. "He's a little high-strung."

There was a knock on the door and then it swung open. Roger entered, followed by Ted and Evanson. Roger was grinning.

"Ted has decided to give us a firm one-year order," said Roger. "520 wind-up alarm clocks per month."

"Well, that's wonderful," said Frank, as he introduced himself to Ted.

"I've never seen a company turn around so fast," said Ted. "I have a great deal of trust that you will meet your quality and delivery commitments. I know you're not done improving yet, but this is the way a manufacturing company should be run. If you do well this year, we'll stay with you indefinitely."

They heard angry voices and a stampede of footsteps coming down the hall. Mortimer appeared in the doorway and glared menacingly at Tom. "I've had it with you, pal," he said. "I'm selling this company out from under you whether you like it or not."

"Wait just a minute, Mortimer," said Frank. "Seems to me our friends here have got themselves back on their feet. We're not going to sell out a home-town family business because of a few late debt payments."

"You've messed with me for the last time," Mortimer said to Frank. "I quit. Acme offered me a job three months ago, and I'm taking it."

"We're not going to need you," said the heavyset bald man behind Mortimer. "We thought you'd get this company for us, and all you've done is waste our time." He turned and left, followed by six lawyers and accountants in dark suits.

"Good luck on your job hunt, Mortimer," said Frank.

There is a critical transition as you move your organization through the lean transformation, a point when managers must become coaches rather than tyrants and employees become proactive. This transition is the key to a self-sustaining organization.

- From *Lean Thinking*, by Womack and Jones

The Next Step

Wednesday, May 11   10:00 a.m.   Tom's Office

"Are we ready for next week's events?" asked Roger.

"Yes," said Tom. "Sam's leading the molding team and I'm leading the receiving/inspection/accounts payable team*. What are you doing next week?"

"I'll be available if you need me," said Roger, "but I'll mostly be working with wind-up assembly and case fabrication."

---

* The kaizen event methodology can be used just as effectively in the office as it can on the shop floor. Many of the lean ingredients also apply. For example, it is often possible to combine receiving, inspection and accounts payable into one work cell and cross train the team members.

"I thought we were done with those areas," said Tom.

"Not even close," said Roger. "We got the kaizen team members excited about making improvements, but now we've got to keep that going."

"Why would it stop?"

"Tom, let me ask you something," said Roger. "During the kaizen events, who made the decisions and came up with the ideas that got us the results we were after?"

"The hourly employees on the team," said Tom. "You told the salaried people we were supposed to keep our sub-teams focused on their goals and help them understand the lean ingredients."

"Did the members of your sub-team seem to be enjoying themselves?"

"Sure."

"Why?"

"They liked the challenge," said Tom. "They liked using their brains and creativity to solve problems. We were counting on them and they responded."

"Okay," said Roger. "Who normally does all the decision making and problem solving on the shop floor?"

"The supervisors."

"Might that cause a problem after a kaizen event?"

"Oh, I see," said Tom. "After getting a taste of a kaizen event, the kaizen team members might not be too excited about going back to being treated like children by their supervisors."

"Right," said Roger. "We treated them like adults, and they responded like adults. We trusted them to save our company, and now we've sent them back onto the shop floor where they have to punch a time clock to prove they got to work on time."

"So what are you going to do about it?"

"First I'm going to get a crowbar and rip the time clock out of the wall," said Roger. "I'll leave the hole there as a reminder, and put a note next to it telling people that from now on we're trusting them to do the right thing. If they want vacation, they'll clear it with their team. If they're sick, they'll call their team as soon as they know. They'll take care of their own time cards and turn them in on Mondays."

"Interesting approach," said Tom. "Then what?"

"The time clock is one example of treating employees like children," said Roger. "If you treat people like children, they won't give you their brainpower, and that's what we need every day, not only during a kaizen event."

"So how are we going to make that happen here?"

"I'm going to teach our supervisors how to be

coaches," said Roger. "I want them to do what you did during our kaizen events. They need to learn how to help their people develop improvement ideas to meet their quality, cost, and delivery goals."

"Sounds a lot like a kaizen event."

"Yes," said Roger. "I want every employee to be thinking of improvement ideas all the time and spending a little of each day implementing them."

"If our employees don't have to worry about being laid off due to productivity improvements, then they'd probably want to help increase their job security," said Tom.

"Yes," said Roger, "and we may even want to consider a company-wide bonus plan to share the gains. That would motivate them even more to help make improvements."

"Then that's what we'll do."

Toyota uses kaizen workshops first and foremost as a human development tool. Yes, they also improved the worksite and, yes, they removed waste, but you realize the real power of kaizen when all employees are applying it in their work every day. Events can instill that kind of thinking.

- John Y. Shook in *Becoming Lean*, edited by Jeffrey K. Liker.

Epilogue

From the May 6 edition of the Ft. Wayne Chronicle, Business Section:

Accurate Clock Company Expanding
Growth Will Create 75 New Jobs

A little over a year ago, after 14 straight months of losses, Accurate Clock Company lost their contract to supply West-Mart with wall clocks, and with it nearly half of their revenue. The local business community pronounced the company dead.

That's history. After one of the most dramatic turnarounds ever witnessed by this reporter, Accurate Clock Company today announced their plan to build an additional plant that will allow them to double their capacity and add 75 additional employees.

Said Roger Dominick, president of Accurate Clock, "We've changed the way we do business by eliminating wasteful work. Our employees have implemented thousands of improvements over the past year and as a result, our quality, cost, and delivery are the best in our industry. We've taken business away from our competitors, and have actually driven several of them out of our market, including two in Southeast Asia. Our goal is to completely dominate the clock industry."

Mr. C. Thomas Langdon IV, CEO of Accurate Clock Company, was unavailable for comment. He is currently on his honeymoon in Bermuda. He married the former Ms. Maria Vasquez, who accepted a position as chief financial officer at Accurate Clock in June of last year.

A Salesman's Encounter with Kaizen

"Curtis, let me make this real simple for you. Either our order is on our loading dock by 8AM Thursday or you don't have an order. You did the same thing to us with the last order, but this time it's in your hands. Either the order is here and you have a customer or it's not and you don't. I'm okay with it either way. Don't call back with any more excuses." Click.

Curt was numb. "Whew, where is Marty when I need him?" He knew Marty, VP Sales, was traveling, but he dialed his cell phone and left a desperate message anyway about the Shantilly order. Curt dialed the company receptionist "Hello, Mary? Can you connect me with someone over there in manufacturing or shipping? I'm not sure who to even ask for, but this is a four alarm emergency and I need to speak to someone." He had to have this order to have any chance to make his sales

234

quota this year. If he was going to lose it at least he wanted to make sure Marty knew it wasn't his fault.

*   *   *

"Who is this?" Cindy demanded.

"Hi, this is Curtis from Corporate Sales." Before Cindy could respond Curt blurted out, "I have a customer emergency and I need to speak to someone ASAP. Can I come over and talk to you about it right away?"

"In an hour I can give you 10 minutes, come to my office in Building 3, first floor," she said.

*   *   *

Cindy held up her hand to stop Curt when he appeared at her door. She left the meeting going on in her office, and met him in the hall.

Curt told Cindy about the conversation he had with Roy and the consequences. "I'm in charge of scheduling," Cindy said. Curt felt energized; he hit the jackpot. "We've been working on some things to speed up delivery time, but I don't think they're done yet," she continued. "I have to speak to some folks before I can give you a firm commitment to tell Roy," Cindy said.

Just then, Danny and Will were walking down the shop floor center aisle. They were the Stamping and Chip Board Assembly Supervisors and it looked they were getting back from a very late lunch.

"Hmm, are any orders going out today?" Curt wondered.

"Hey guys, you got a minute?" Cindy asked. She explained Curt's situation. "I know you guys were tied up in a kaizen event all week. How is it going?"

"We're just getting back from our kaizen celebration lunch," Will said.

"What a week! We reduced our changeover time from 8 hours to 1 hour in stamping and we still have further improvements to make," Danny said.

"Chip Board Assembly is a now part of work cells and the products fly through," Will added.

"There's your answer Curt. These two departments were the last to go through kaizen and the toughest to speed up, and you heard the results. Call Roy back this afternoon and tell him we're still his supplier," Cindy said proudly.

*   *   *

"Your order will be there by 8AM Thursday Roy, no problem," Curtis said.

"Thanks for the call Curt, and please don't take what I said personally the other day. You guys didn't leave me any choice; I'd have to go with another vendor if I can't rely on you. Thanks again for the follow up, and keep in touch," Roy said.

*　　　*　　　*

The following Monday, in the sales meeting, Marty asked, "So Curt, what are you going to do this week that will increase your chances of making your quota?" He asked this same question at the end of every sales meeting. This time Curt had a good answer.

"Well, from now on I'm going to stay in close contact with our manufacturing Supervisors. If not for them I would have lost the Shantilly order and any chance to make quota this year. I don't know how they did it, but there were able to ship an order that used to take 21 days and shipped it in 12 days. I always knew we were the quality leader, but when we combine that with the shortest delivery time we are unbeatable. We're now the leader and I am going to use it at every opportunity to win more business." Curt said.

"That's a great idea. Each week I'll ask one of the Supervisors to join us for our Monday meeting to update us on improvements they are making, and discuss how we might use them to win more business for the entire team." Marty said.

"Besides, I want to find out more about this "kaizen" stuff, I think it will be a big seller." Curt said.

Somehow we have to get past the idea that all we have to do is join hands in a circle and sing Kum Bah Yah *and say, "We're moving to teams." It's just not that easy, and anyone who has ever tried it knows it.*

- Michael Shrage, author of *No More Teams*

Bonus Article

How To Get Employees To Think Of Improvement Ideas

Are your first line supervisors fire-fighters or leaders? Fire-fighters (also known as traditional supervisors) do little more than administrative work and conflict resolution. Leaders (also known as coaches) motivate their people to improve quality, cost, and speed for the benefit of everyone in the company.

An important part of being a coach is leading individual contributors to think of and implement improvement ideas, and an experienced coach does it naturally. For a traditional supervisor trying to make the transition, however, it can be tough to get started. In this article we'll give you the top 12 tips to help employees think of improvement ideas. In our next article we'll talk about how to get them to actually implement those ideas.

The Top 12 Idea-Generating Tips:

1. Give your team a reason to think of improvement ideas. Many people believe "if it ain't broke, don't fix it." You need to establish measures to show "it's broke." Figure out a way to measure your quality, cost, and speed, find out how your department is currently performing, and set a six-month improvement goal. Put it all on a big graph and post it in a place where your team can't miss it. For help, go to www.everestcg.com/JTT_measures.htm.

2. If you are a traditional supervisor, you might think your job is to solve problems. Stop it! Right now you are probably better at solving problems than your direct reports, but they outnumber you, and if you can get them to apply their experience and brain power to solving problems and making improvements, they will soon blow you away. But they won't start until you stop. It will be frustrating for a while, but once they know you are counting on them to do the thinking, they'll make your life a lot easier.

3. Problems and improvement opportunities are like elephants. How do you eat an elephant? One bite at a time. Ask for and accept small ideas at first. Your people don't need to change the world the first week, and doing little things is less scary than doing big things. Lots of small ideas will add up to lots of improvement in a short amount of time.

4. Make sure that all your direct reports are thinking of ideas, and that everyone on your team knows it. If one person is not participating, other people will think they don't have to. Handle the non-participators behind closed doors, coach them to think of any idea at all (see #3 above), then make sure the team knows you've dealt with the situation.

5. Teach your people the concept of value-added work. Value is added during manufacturing only when material is being physically changed or assembled. In the office, value is added only if the work involves a mathematical manipulation of numbers, composing or editing of textual documents, or analysis of numbers, words, or ideas to produce knowledge. Ask your individual contributors to identify non-value-added work and think of ideas to get rid of it. The most common forms of non-value-added work are waiting and moving. For help go to www.everestcg.com/kaizen_lean_VA.htm.

6. Ask your people what annoys them most in the course of their work. Annoyances are non-value-added. Thinking about how to get rid of annoyances can lead to improvement ideas, and shows them you care about their morale and comfort, not just the bottom line.

7. Improvement ideas don't always have to work. That's why they are called ideas, not solutions. Just thinking of an idea to try (even if you don't know whether it will work) is fine. If you have this attitude, your people will be much more willing to think of ideas.

8. Let your employees try ideas you don't agree with but that won't hurt too much either. More than any other leadership behavior, this shows you respect their ideas and think of them as your equal in their ability to think. If their idea works, they will feel great and be motivated to think of more ideas. If their idea doesn't work, they'll learn from it. Remember: when your people cause a change, they feel more ownership for the results of your department, and will try harder to make sure those results are good.

9. Discuss with your people the last few production problems your department experienced. Ask why those problems occurred and what they can do to eliminate them in the future. Whatever they say, agree with them and ask them for concrete ideas they can implement to solve the problem.

10. Have your direct reports talk to internal suppliers or customers, or to other departments like engineering, product development, or marketing. Have your people ask the other departments what they think could be improved about your department.

11. Teach your people about 5S (Sort, Straighten, Scrub, Schedule, Score). This workplace organization strategy applies in every work environment, and will result in hundreds of improvement ideas. To learn more, go to www.everestcg.com/Kaizen_Lean_5S.htm.

12. Get your people crossed-trained to do more than their individual job. Have them learn every job necessary to make/deliver the complete product/service produced by your department. If possible, have them learn jobs outside their department. The more they know about your company, the more problems and opportunities they will see. After all, your entire firm is one big business process with lots of room for improvement.

Thinking of improvement ideas will be a challenge for your team members, but the challenge will make work more rewarding for them. Use the twelve ideas above to challenge them successfully.

If you liked this article, call 1-888-910-8326 to order your copy of The Journey To Teams, *a book by Mike Regan (with Mark Slattery) describing the new way to implement teams.*

Additional Resources

- Experienced consultants and trainers are available from Everest Consulting Group, Inc. to lead kaizen events and to teach you how to lead them yourself. We also deliver comprehensive training, consulting and keynote speeches in the areas of kaizen, lean manufacturing, and team building. Call 888-910-8326 for more information.

- To receive our KAIZEN or TEAM ANSWERS e-mail newsletters, please visit www.everestcg.com to subscribe.

- Explore our web site at www.everestcg.com.